Library of
Davidson College

MICHIGAN BUSINESS
REPORTS *New Series* NUMBER 1

The Tool and Die Industry
Problems and Prospects

HAROLD E. ARNETT
DONALD N. SMITH

A publication of the
Division of Research
Graduate School of Business Administration
The University of Michigan
Ann Arbor, Michigan

338.4
A948t

Copyright© 1975
by
The University of Michigan

76-9719
All rights reserved
Printed in the United States of America
ISBN–0–87712–173–7

CONTENTS

	Preface	xi
I.	Approach of the Research and General Observations about the Industry	1
II.	The Nature of and Significant Trends in the Tool and Die Industry	6
III.	The Nature of Costs and Their Classifications for Decision Making	19
IV.	Pricing Policies	28
V.	Market Characteristics and Prospects of the Tool and Die Industry	43
VI.	Economics of Independent Tool and Die Shops and the Make-or-Buy Decision	55
VII.	Should Captive Shops Be Utilized?	67
VIII.	Conclusions and Recommendations	76
	Appendix A: Statistical Tables	83
	Appendix B: Make-or-Buy Programs of the Big Four	96

TABLES

1. Responses to the Questionnaire 4

2. Total Shipments for Tool and Die Products, 1972 9

3. General Statistics on Establishments in the Special Dies and Tools Industry (SIC 3544) 1972, 1967, 1963, 1958, and 1954 9

4. Selected Statistics for Special Dies and Tools by Employment Size of Establishments, 1972 10

5. Employment Changes in Various Categories 16

6. Key Milestones in Two Decades of Expansion of Captive Shop Toolmaking 51

7. Direct Labor and Fringe Benefit Costs, Independent and Captive Shops 62

A.1 Value of Shipments and Employment in the Tool and Die Industry (SIC 3544) for 1947–72 83

A.2 Value Added by Manufacture for the Tool and Die Industry (SIC 3544) and Gross National Product 1947–72 84

A.3 Converting Gross National Product and Value Added by Manufacture for Special Dies and Tools to Constant Dollars and Percentage of Growth: 1947–72 85

A.4 Regional Distribution of Tool and Die Establishments 86

Tables *cont'd.*

A.5 Leading States According to Value Added by Manufacture in
the Tool and Die Industry: 1972, 1967, 1963, and 1958   87

A.6 Distribution of Employment by Firm Size
in the Michigan Tool and Die Industry   88

A.7 Value Added per Production Worker Man-Hour
for Tool and Die Selected Customers' Industries   89

A.8 Bench-Mark Data for Table A.7   90

A.9 Portion of U.S. Tool and Die Industry in Michigan
1972, 1967, 1963, and 1958   95

FIGURES

1. Inverse Relation between Tooling Size and Suppliers in the Tool and Die Industry — 12
2. Decline in Toolmaking Activity of Detroit Tooling Association Companies — 15
3. Scatter Diagrams of Past Periods' Costs and Volume — 23
4. Cost and Volume as Interrelated with Revenue and Profit — 25
5. Alternative View of the Relation of Costs, Volume, and Profit — 26
6. Simple Profit–Volume Relationship Reflecting Sales and Cost Patterns — 26
7. Annual Expenditures for Special Tools and Dies for the Big Four Auto Makers — 45
8. Variations in Model Tooling as Measured by Amount of Iron Cast for Body Dies — 46
9. Relation of Cyclical Tool-and-Die Demand to Captive Shop Expansion — 53
10. Diemaker Wage Rates at Captive and Independent Shops — 61

PREFACE

In December 1974 we undertook a project sponsored by the Detroit Tooling Association and the National Tool, Die, and Precision Machining Association to look into the problems facing the industry, particularly those peculiar to the large automotive-body die shops in the Detroit area. Our objective was to look for areas which could be improved to better the industry. The undertaking has been interesting and challenging.

We were not instructed to make a detailed and exhaustive analysis of the industry's problems. Such analysis would require much more intensive study than was possible as a part of this project. Our purpose was to bring the problems into clearer focus, to convince the members of the industry that these problems really do exist and need attention, and to recommend broad paths or guidelines to follow in searching for solutions. In so doing, we have made heavy use of two other studies: (1) William A. Paton and Robert L. Dixon, *Make-or-Buy Decisions in Tooling for Mass Production,* Michigan Business Reports, Number 35 (Ann Arbor: Bureau of Business Research, Graduate School of Business Administration, The University of Michigan, 1961), a study concerned with many of the same problems we studied—one of our objectives was to update and expand that study; and (2) R. Lee Brummet and Harold E. Arnett, *Business and Economic Evaluation of the Metal Finishing Industry* (Michigan Business Reports, Number 52, by the same publisher, 1967), a study of another industry having many of the same problems as the tool and die industry. We are grateful to those authors and the Bureau of Business Research for allowing us to make such liberal use of the information in those studies.

We recommend that those who have only a casual knowledge of the tool and die industry read the entire report before considering the conclusions and recommendations in Chapter 8. The first seven chapters present the evidence and analyses in a way that leads logically, we hope, to those conclusions. Read or quoted by themselves the recommendations and conclusions may be subject to misinterpretation by the uninitiated. Those who are knowledgeable about the industry may wish to go directly to Chapter 8, although we recommend they also read the entire report first, particularly Chapters 6 and 7.

We wish to thank the associations who sponsored this project and their representatives and members who gave such enthusiastic support and assistance throughout, particularly those who responded to the rather formidable questionnaire and those who welcomed us into their plants to study their operations and problems.

We also express our thanks to Robert Stricof, candidate for the M.B.A. degree from the Graduate School of Business Administration, University of Michigan, for his assistance in collecting and analyzing information during the early stages of the project.

Harold E. Arnett
Donald N. Smith

Ann Arbor, Michigan
August 1975

I

APPROACH OF THE RESEARCH AND GENERAL OBSERVATIONS ABOUT THE INDUSTRY

Approach of the Research and Questions to be Answered

This study analyzes the structure of the tool and die industry, incorporating observations about its sales, manpower, cost trends, and competitive problems. We investigated independent and captive shops to determine their relative strengths and weaknesses. Specifically, we sought answers to the following questions:

1. What are the basic characteristics and trends of the industry including the number of firms, and their size, structure, location, employment, payrolls, and value of shipments?
2. What are the key competitive problems facing independent tool and die firms?
3. How does the cost structure of independent firms influence their competitive position?
4. How do independent firms in the industry set prices, and do these practices influence competitive position?
5. What are the primary economic and business justifications for the continued existence and growth of independent tool and die firms?

A corollary to these is: Is it in the economic interest of major manufacturing concerns to continue to produce substantial amounts of their tooling requirements, or should there be a leveling off or retrenchment from the present volume of self-service? Answers, of course, depend on the answers to several subsidiary questions: (1) What criteria do mass production industries use in tooling make-or-buy decisions? (2) What functions are more economically performed by the smaller, independent companies? (3) What functions are more economically performed by the large captive shops?

2 / *The Tool and Die Industry*

To obtain data regarding these questions, we analyzed the general characteristics of the industry, gathering information from published sources, association records, and selected field interviews. We interviewed the owner-managers of independent tool and die firms to secure the benefit of their experience and to learn about their operations. In all, we visited some twenty firms located in and around Detroit, including three captive shops. In selecting these firms, we attempted to cover the various types of tool and die processes—captive and independent, unionized and open. This cross-section of operations gave us an opportunity to observe the major types of activities and to determine the major problems of the industry as a whole. At the same time, we sent a comprehensive questionnaire to 100 firms—half went to members of the Detroit Tooling Association; the other half went to members of the National Tool, Die, and Precision Machining Association, throughout the United States. Since tool and die shops in Detroit are experiencing the most difficulties in the industry at present, we felt it was appropriate to send 50 percent of the questionnaires to this group. The fifty sent to the national group helped insure that we would also observe and give adequate consideration to various geographical factors and problems peculiar to them. There were thirty-three usable responses to this questionnaire. The answers received were very useful in pinpointing some of the major problems facing tool and die firms and in indicating some of the weaknesses which the industry needs to overcome if it is to maintain and improve its position.

Basic Technical Aspects of the Industry

The economical production of durable commodities, such as vehicles, electrical appliances, office machinery, and farm equipment, requires they be manufactured in mass quantities. Such production is possible only through the heavy utilization of "tooling." Generally, tooling consists of dies, jigs, fixtures, molds, gages, and specially designed machines. Combined with adequate managerial know-how, a trained labor force, and production-line facilities and techniques, these elements constitute modern mass production. This study is broadly concerned with the tooling of:

    a) *Dies.* A die set consists of a pair of cutting or shaping tools which, when moved toward each other, produce a certain desired form in, or impress a desired device on, an object or surface by pressure or a sharp force. The term "die" may also refer to one of the basic die set members; the "punch" is the other.

    b) *Jigs and fixtures.* Jigs are devices for supporting the workpiece and for guiding the cutting tool of the machine tool during processing. Fixtures are of several types, but the more typical ones support or hold in place a workpiece during its processing; others are used in assembly and checking operations. In general, jigs and fixtures may do all or some of the following operations: locate, clamp or support a workpiece, and guide a tool. Fixtures normally are not involved in the latter operation.

c) *Mold.* A device that forms parts as molten metal, rubber, plastic, or comparable material is fed into it.

d) *Gage.* An instrument used to determine whether a given part dimension is within specified tolerance limits.

e) *Special machines.* Nonstandard machine tools, usually used for metalworking operations, and mostly of a metal-removal type.

The term "tool and die" has evolved through trade usage to describe the sector of the economy which includes toolmaking companies. Its limits are difficult to define rigorously. In addition to those companies producing the tools and dies defined above, a manufacturer performing precision machining operations or closely related activities can associate his organization with the tool and die industry.

For some purposes, strict definitions of the technical aspects of tool and die work, such as those which appear in the U.S. Office of Management and Budget's definitions in the *Standard Industrial Classification Manual: 1972,* may be useful. Since this study is basically concerned with a financial and economic analysis of the industry, however, strict applications of technical definitions are not desirable. All types of companies share many of the same economic and financial problems, as well as the competitive challenges stemming from the make-or-buy decisions of mass production industries. We focus on that phase of tool-and-die operations associated with automotive requirements. Such work accounts for about one-third of the industry's $2.0 –$2.5 billion output.

Specific Industry and Firm Problems

In response to questions "What do you consider to be the major problems of your firm?" and "What do you consider to be the major problems of the tool and die industry in general?" respondents to the questionnaire answered as shown in Table 1.

Although these problems will be considered in greater detail in later chapters, a few observations are warranted here. Table 1 summarizes the responses in the thirty three questionnaires returned by firms throughout the country. In the Detroit area, however, the reduction of demand brought on by the increase in captive shops represents the major problem cited by respondents. In the summary, like items were combined to form the six classifications in the table, thereby losing the replies' individuality. To counterbalance that, a few of the respondents' individual statements regarding their problems follow. They are revealing.

4 / The Tool and Die Industry

Table 1
RESPONSES TO THE QUESTIONNAIRE

	Times Cited	
Major Problems	Number	Percentage of Total
Of the firm		
1. Unfair competition—pricing below costs because of lack of knowledge of cost structure (also includes depressed pricing brought on by excess productive capacity).	15	27
2. Growth of captive shops, reducing demand for outside work and causing excess productive capacity (includes loss of independent firm employees to captive shops).	14	25
3. Lack of experienced and competent help (includes poor attitude and motivation of employees).	12	22
4. Increasing costs of production (materials, labor, machinery, etc.) and inability to pass these on to the customer (includes both inflationary effects and upward pressure exerted by unionized captive shops on labor costs).	6	11
5. Inconsistent and repressive governmental policies (includes high taxes which destroy business incentive).	5	9
6. Depressed conditions of U.S. and world economies.	3	6
Total	55	100

Continued

"The union is our problem; there is constant antagonism."

"Government intrusion into the automotive industry created most of our problems."

"Many customers acquire five quotations. Often one of these is too low and costs us the job."

"Burdensome taxes. Government laws, regulations, and reports governing small businesses are expensive and time-consuming to adhere to, and repressive in their effects on operations."

"Build-up of captive die capacity which decreases available large panel die work. This results in wide swings of activity with increasing low activity periods and low prices."

Table 1, *cont.*

Major Problems	Times Cited	
	Number	Percentage of Total
Of the industry in general		
1. Unfair competition—pricing below costs because of lack of knowledge of cost structure (also includes depressed pricing brought on by excess productive capacity and pressure from big business for price reductions).	16	37
2. Lack of experienced and competent help (includes poor attitude and motivation on the part of employees).	8	19
3. Growth of captive shops, reducing demand for outside work and causing excess productive capacity (includes loss of independents' employees to captive shops).	7	16
4. Increasing costs of production (materials, labor, machinery, etc.) and inability to pass these on to the customer (includes both inflationary effects and upward pressure exerted by unionized captive shops on labor costs).	5	12
5. Tight delivery schedules demanded by the customer with insufficient lead time allowed.	4	9
6. Inconsistent and repressive governmental policies (including high taxes which destroy business incentive).	3	7
Total	43	100

"Availability of skilled manpower. Captive shops do not do their share of training. Training is hampered by increasing low activity periods."

"Captive shop piracy of our employees—based on higher wages and benefits due to union pressure."

"A need for leveling of the work load curve. Stylists and product people use all the lead time and then expect tools to be built in short order."

"We are gradually being frozen out by the large manufacturing plants (captive shops)."

"We cannot command high enough prices for our jobs due to fiercely competitive pricing. Not enough work is being let out by the Big Four auto makers."

"Price-cutting tactics in the industry. People not knowing their costs."

II

THE NATURE OF AND SIGNIFICANT TRENDS IN THE TOOL AND DIE INDUSTRY

While mass production is made possible by tooling, the principal tools themselves cannot be mass produced. Tool making, and especially mold and diemaking, is one of the few activities connected with modern large-scale industry in which there has not been a general substitution of machinery for basic skills. These tools are custom-made, one-at-a-time by skilled artisans who patiently and precisely machine, finish, and construct the complicated devices. Only one die, or set of dies, is needed for the manufacture of many thousands, and sometimes millions, of automobile fenders or hoods of a given design.[1] This same principle also is generally applicable to mold and, to a lesser degree, jib and fixture requirements. The one-of-a-kind characteristic of the tooling industry accounts for enormous differences in management and capitalization strategies, and the skills, machinery, and technology amenable to toolmaking and mass production.

Tool and die shops can be broadly classified into two groups: independent (job) and captive. Many of their functions are alike. Both build tooling, and both "try out" the tooling to assure its satisfactory performance on the production lines. Depending on the circumstances, either class of shop may modify a tool in the tryout and early production-run period to accommodate structural improvements. Both classes may, and are able to, perform the final adjustment and adaptation of the tool in its production-line setting, and both may maintain the tools during the course of their utilization in production runs. Final adjustments and adaptation, and production maintenance, however, have been performed mainly by captive shops adjacent to the production lines in which the tools will be used. The design function is also performed by both, but generally rests with the engineering design staff of the organizations owning the captive shops. Thus, although there are many similarities, there are also significant differences between independent and captive shops.

Independent Tool and Die Shops

Most job shops began and have continued as family operations. If they are not family proprietorships, they are generally either partnerships or closely held corporations. The people involved (owners, management) tend to be strong, free-enterprising, and very independent. Most job-shop owners and managers are self-made men—highly motivated, competitive, hard working, and strongly individualistic. While these characteristics have contributed much to the growth and perpetuation of the industry, they have also had some unfavorable effects. For instance, some job shop managers have been reluctant to discuss their problems with others in the industry. They have been hesitant to seek solutions through exchange of cost information and pricing practices and only infrequently do they look for common ground with competitors on matters of mutual interest. They have been hesitant, also, to recognize that they can engage in fruitful associations with competitors without losing independence or divulging confidential information. Such reticence must be overcome if solutions are to be found for problems common to shops in the industry.

As service organizations

Nationally the tool and die industry consists of approximately 6,500 companies. Approximately half of the total U.S. plants are in Michigan and the states adjoining it. Nearly a fifth of the U.S. plants operate in Michigan, employing about 30 percent of the national work force.

Unlike other segments of American industry, the tool-and-die sector is not characterized by industrial giants. Nearly all the companies are small business enterprises. In 1972, 81 percent of these plants had fewer than twenty employees; this has remained almost constant over the past several years. Out of the 6,616 tool and die establishments operating in 1972, 4,014 had fewer than ten employees; only 347 had more than fifty employees. Independent companies producing automotive-body tooling tend to be larger than the industry average, ranging in size from thirty to 150 workers. In total, independent shops represent a significant force in the economy. (Appendix A contains significant data regarding value of shipments, employment, value added by manufacture, etc.)

The final product of the tool and die company demonstrates only a portion of its total activity. The overall scope of a firm's operation is determined by the point at which the service of an order is initiated. In some instances complete information on every detail of the tooling and all phases of its manufacture is provided by the customer. Conversely, there are circumstances in which the company is consulted prior to the finalization of the production plans. In this event, the tooling company evaluates the feasibility of the buyer's plans, perhaps modifies them, then designs and builds the tools, and occasionally even sets up all phases of the production-line tryout and operation.

The difference between these extremes, ranging from minimum to maximum utilization of engineering skills, is extremely important. At the former extreme, the tool and die company is performing a strictly mechanical machining service. In the latter situation, however, the shop also acts in the capacity of production engineering consultant, and the indispensability of the service provided is markedly increased. Thus, although two companies may seem to have identical end products, their character may differ substantially. This difference is usually referred to as the extent to which a company does its own design and development.

The word design does not by itself clearly define the firm's overall range of activities, however. Under certain conditions design signifies that the customer has provided a complete set of drawings and often models of the finished tooling that the tool-and-die shop will build. In this case the supplier must decide on the intermediate steps in the production process. At other times a customer may provide only drafts or models of the component he wishes to manufacture, and the tooling company must completely design the necessary tools. In most instances, independent tool and die shops do not produce a *final* product. They are service organizations, characterized by product or service specializations, which produce tooling other firms use to produce and sell products.

Specialization and diversification

Particular demands of regional marketplaces and the mechanical skills of the management and work force profoundly influence the type of specialization a company selects. Because they are located near the automotive design and engineering centers, the tooling companies operating in Michigan, and to a lesser extent in Ohio, Indiana, and Illinois, are heavily committed to automotive needs. By contrast, West Coast tooling companies adapt to aerospace requirements.

The industry is versatile and is able to adapt readily to new demands as they arise. For example, in addition to the dictates of market requirements, the relatively large amount of capital needed to equip an all-purpose tooling shop indirectly forces independent shops to specialize. The facilities of individual shops therefore are tailored to a particular tool size, such as dies weighing less than one ton, or dies over ten tons, die-casting dies, or production of large or small molds, gages, etc. Other companies emphasize their experimental and research equipment, their tryout or short-run production facilities. A specialization growing along with the use of plastics is the mold-making facility, requiring precision boring, milling, and layout equipment. The Chicago area has been an important mold-producing center. Thus, specialization is not only an effect of the facility's makeup, but also is closely tied to the nearby demands of the marketplace and to the mechanical skills of the company originators.

Since independent shops are service groups performing a large variety of operations for their customers, they have their own distinct advantages and problems. The success of an independent shop depends on the degree of its customers' activity. Consequently, no matter how efficient and well-managed a job shop may be, it cannot survive unless its customers generally enjoy successful operations *and* buy from the independents rather than manufacturing their own tooling needs. This dependency on other industries presents the problem of maintaining a

Table 2

TOTAL SHIPMENTS FOR TOOL AND DIE PRODUCTS, 1972

Product	Price in Millions of Dollars	Percentage of Total
Jigs and fixtures	318.5	11.7
Dies	1,201.6	44.3
Industrial molds	640.3	23.6
Other	553.1	20.4
TOTAL	2,713.5	100.0

Source: U.S., Dept. of Commerce, Bureau of the Census, *Census of Manufactures: 1972* (Washington, D.C.: Government Printing Office, 1972), MC 72(2)–35C.

Table 3

GENERAL STATISTICS ON ESTABLISHMENTS IN THE SPECIAL DIES AND TOOLS INDUSTRY (SIC 3544) 1972, 1967, 1963, 1958 and 1954

Year	Total of Firms	Firms with 20 Employees or More	Percentage of Firms with Fewer than 20 Employees
1972	6,616	1,267	80.8
1967	6,615	1,531	76.9
1963	5,896	1,133	88.8
1958	5,745	994	82.7
1954	5,209	1,054	79.8

Source: U.S., Department of Commerce, Bureau of the Census, *Census of Manufactures: 1972*, MC72(2)–35C; *Census of Manufactures: 1967*, p. 35C–11; *Census of Manufactures: 1963*, p. 35C–7; *Census of Manufactures: 1958*, p. 35C–6, and *Census of Manufactures: 1954*, p. 35B–4 for adjusted 1954 figures. (See first footnote in Appendix A, Table A–1, for description of adjustment.)

Table 4
SELECTED STATISTICS FOR SPECIAL DIES AND TOOLS BY EMPLOYMENT SIZE OF ESTABLISHMENTS, 1972

Item	Total Establishments	Under 10	10–49	Number of Employees Over 50	Over 100	Over 250	Over 500
Number of establishments	6,616	4,014	2,255	347	123	16	7
Percentage of total	100	60.7	34.1	5.2	1.9	0.2	0.1
Value of shipments (Millions of dollars)	2,426.7	359.0	1,074.6	993.1	629.2	217.5	164.7
Percentage of total	100	14.8	44.3	40.9	25.9	9.0	6.8

Source: Derived from U.S., Department of Commerce, Bureau of the Census, *Census of Manufactures: 1972*, MC72(2)–35C.

sustained high volume of activity. This problem can be reduced, and, in some cases, is being reduced in two ways.

First, independent companies frequently undertake projects which they are not, individually, equipped to perform fully. This is made possible through long-standing and highly efficient subcontracting, or contracting on a cooperative basis. In effect this means the facilities of an entire area are available to the customer whenever a tool requirement extends beyond the capability of an individual shop. Several locales have effective trade associations that formally coordinate available skills, machinery, and facilities. When a certain requirement exceeds the available resources in a particular area, the National Tool, Die, and Precision Machining Association's information network can be tapped to indentify available facilities in nearby areas. Special precision boring operations, very large-scale copy milling, unique heat treating, electrical discharge machining, etc., may be subcontracted to specializing shops.

Each independent company is, therefore, freed of the expense of maintaining the complete complement of expensive machinery required for all processes, some of which may be needed by only one company for a few weeks each year. By cooperative contracting, a shop may be able to bid on work in markets that would be outside its capabilities otherwise. As a group, the independent companies in the Detroit area have been termed the "largest tool, die, and special machine shop in the world." Their combined capability, plus their ability to work together in speedily tooling-up production lines for World War II, earned them the international reputation "arsenal of defense."

Second, some independent firms go even further, developing a sizable number of different types of customers rather than depending on one or a few. With such diversification, operating activity may be reasonably stable at a fairly high level throughout the year, despite customers' needs changing subject to wide seasonal and business fluctuations.

In answer to the question, "What percentage of your total business is for nonautomotive customers?" the replies of the companies surveyed ranged from zero to 95 percent. Twelve companies were 10 percent or less nonautomotive and half were 25 percent or less. This high percentage of business tied to the automotive industry is not surprising considering that one-half the questionnaires were sent to firms in the Detroit area. But it does indicate a heavy dependence on the fortunes of one industry and the need to put considerable effort into diversifying output.

There is a significant number of potential, nonautomotive customers both in product lines and markets served. The respondents' list of the types of nonautomotive work they perform showed their work varied from the recreational industry to the aerospace industry, including farm equipment and military ordnance. Products produced varied from tooling for producers of bedsprings to television sets, pumps, and compressors for locomotives. Most respondents indicated they are attempting to diversify; many said they have increased nonautomotive work during the past few years. Still more needs to be done in the area of diversification, however. As will be discussed more fully later, many characteristics of the large-body die shops make it very difficult to diversify, so other solutions must be found for their economic problems.

Many of the operations that produce automotive tooling—particularly the construction of large body tooling—are concentrated in the Detroit area. Ohio, Indiana, northern Illinois, and western New York are also important areas, however. Because an extensive and diverse array of productive equipment and support facilities are required for the economic manufacture of large body tools, auto manufacturers seldom find qualified tool and die sources outside their geographical area, roughly defined as a 300-mile radius extending to the south, east, and west of Detroit. Many body dies used in foreign auto plants have been produced in this region. More recently, foreign toolmaking know-how has developed to a level such that it often competes effectively with American companies in making certain types of tools for American production lines. Low-cost labor has had much to do with the foreign success.

Tooling size capability is an important determinant of economical production. Equipment and skill requirements restrict the number of tooling companies able to produce large dies; their number varies inversely with the tooling size, as shown in Figure 1.

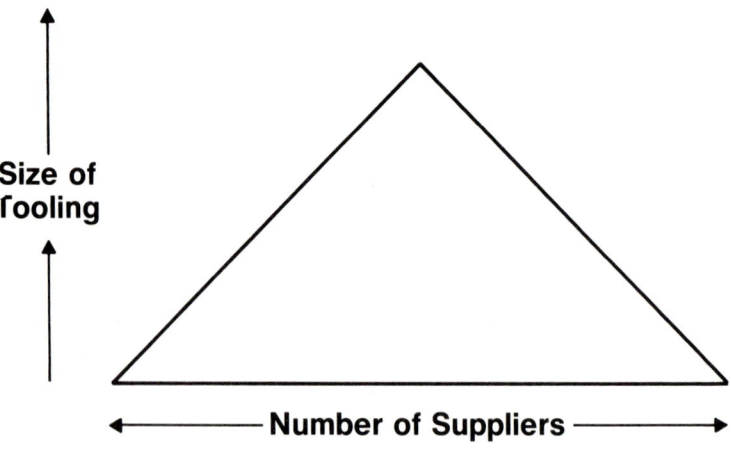

Fig. 1. Inverse relation between tooling size and suppliers in the tool and die industry.

Captive Shops as Tool and Die Makers

A captive shop is one which is owned and operated by and serves a manufacturing company whose primary function is the manufacture of a final product. To such firms tool and die making is only one of many activities necessary to convert raw materials into the product sold to the customer. On the other hand, an independent shop's primary, and perhaps only, function is to produce tooling for others.

Captive shops have grown to undertake all phases of tool and die operations except, usually, producing tools on a large scale for outsiders. In the early 1970s, however, one automotive manufacturer's captive shop constructed a large body die order for another of the Big Four. The stated motivation for this new arrangement was to utilize numerical control die manufacturing methods. This underscored the passing of the decades-old justification for captive shops—the protection of styling secrets.

The minimum purpose of a captive shop is to maintain the dies and other tooling items at the stamping plant. Press lines in automotive stamping plants are highly efficient, producing 500–600 major body panels per hour on lines with automatic handling equipment. At this rate, some two million pieces per year, theoretically, can be produced on a single line. For this reason, press shops periodically change dies when inventories of one body panel have been filled and others have been depleted.[2]

Clearly breakdown of dies during periods of high-speed mass production may constitute an emergency which must be corrected immediately. In such an emergency the opportunity costs of lost production and sales are so high that the costs of repair may be quite secondary in significance. To provide support service on a constant, stand-by basis, mass-production plants have tool and die maintenance shops adjoining the production lines. Paralleling this requirement is the need for on-the-spot facilities to make minor modifications and adjustments to purchased tooling at the time it is introduced into the production lines. While skilled labor is needed for these service activities, the shop that is designed for tool maintenance alone does not require the elaborate equipment necessary for original tooling construction. Thus, while the maintenance shop utilizes hand grinders, drill presses, welders, cranes, etc., it does not need costly planers, boring mills, large copy milling machines, or duplicating equipment.

Because tool maintenance requirements are not continuous (this activity utilizes about two-thirds of the time of a toolmaker in a stamping plant's toolroom), over the years a number of captive shops have sought to level out the utilization of tool maintenance personnel by adding tool construction to the shop's activities. This development, together with its secondary effects, may partially account for the spiraling growth of certain types of captive shops. Die construction, for example, requires relatively expensive duplication equipment, planers, boring mills, shapers, etc., and an extensive array of ancillary facilities. Committing capital to acquire these facilities increases the emphasis on fully utilizing the investment to such an extent that the tool construction tail then wags the tool maintenance dog. This second level of activity, tool construction on a fill-in basis, then leads to a third level in which the captive shop becomes a primary producer rather than a maintainer of tools.

The investment in captive shops possessing a full complement of equipment is enormous compared to that in the typical specialized independent shop. The plant and equipment of larger captive shops is considerably more elaborate than that of the independents. Such captive shops often employ up to 2,000 men, ten to fifteen more than most of the larger independent companies.

Because the attached captive shop has only one customer, seasonality of activity is frequently a serious problem. Some smoothing of seasonal activity is achieved by overlapping programs. For example, in the automotive industry, a large captive shop may alternate between maintaining and changing current-model tooling, construction and tryout of tooling for next year's models, and preliminary machining of templates or tooling components for future models. Even this does not assure total smoothing of activity, since model changes from year to year may be minor, major, or complete, and obviously the changes are not programmed solely for the convenience of the captive tool and die shop.

Variability of activity is another serious problem for those large independent body die shops which are tied almost exclusively to automotive production. For many years they survived fairly well. The profits they earned during heavy utilization of productive capacity in model changeover periods carried them through the leaner months and years in between. Recently, however, work load variations in captive shops led to union demands for some control over subcontracting tooling orders to independent shops that might be used to fill out a captive shop's work load. This is now a strikeable issue, and is a major cause of the problems larger independent body tool and die shops now face.

Major Effects of Captive Shops on Independent Firms

Fifteen out of thirty-three survey respondents indicated that captive shops presented a major problem for their individual firms; eight indicated they were a major problem for the entire industry. A number declared that future implications of captive shops were more serious than the current economic situation, particularly for the large automotive body die shops.

Recently, independent companies that produce automotive body tooling have experienced an accelerating downward trend (see Figure 2). Much of this decrease is a result of the increase in captive shops. A number of tool and die shops have closed, and a substantial reduction in

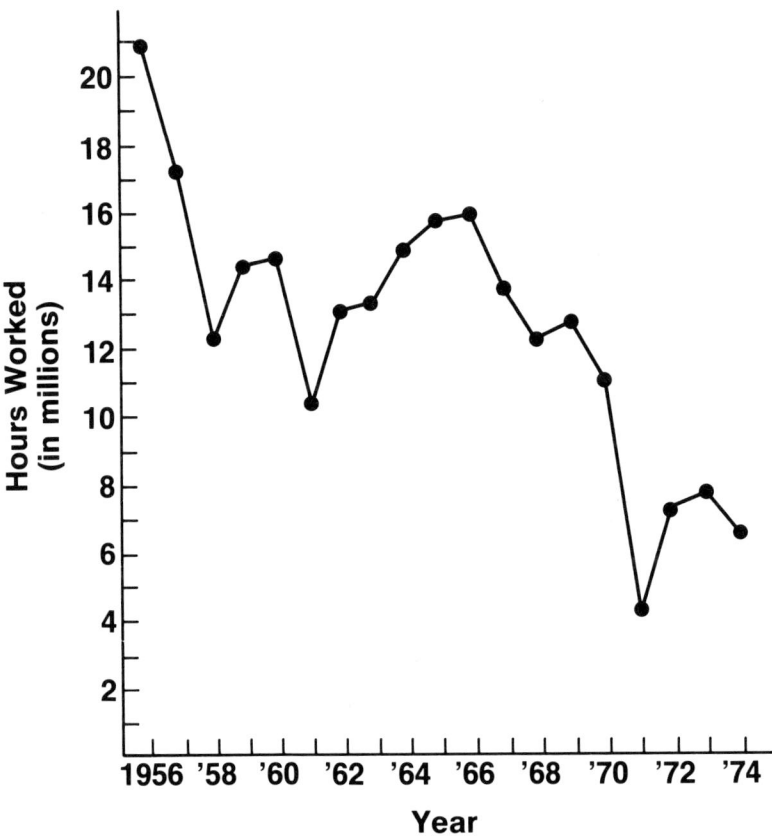

Fig. 2. Decline in toolmaking activity of Detroit Tooling Association companies.

Table 5

EMPLOYMENT CHANGES IN VARIOUS CATEGORIES
(in thousands)

Category	Detroit			Michigan			United States		
	1960	1968	1972	1960	1968	1972	1960	1968	1972
Metalworking machinery employees	33	44	35	1,570	1,822	1,807	7,930	9,386	9,586
Machine tools, metal cutting employees	N.A.	N.A.	N.A.	133	160	166	701	903	900
Machine tools, metal forming employees	N.A.	N.A.	N.A.	45	55	66	317	357	380
Special dies, tools, jigs and fixtures employees	N.A.	N.A.	N.A.	1,093	1,246	1,207	5,795	6,615	6,568

Source: U.S., Bureau of the Census, *Census of Manufactures* and *Annual Survey of Manufactures* (Washington, D.C.: Government Printing Office, 1960, 1968, 1972).

operations has been felt by those remaining. Over the last nineteen years, some seventy-four tool and die shops in the Detroit area have gone out of business. These closings resulted in the loss of approximately 5,900 jobs. Of the twenty largest shops, twelve have closed their doors, eliminating 2,800 jobs, nearly 48 percent of the total. In the eight remaining, employment dropped from 1,332 in January 1970 to 947 in April 1974, and to 359 in early 1975. A number of other statistics are also revealing as Table 5 shows.

Considering only metalworking machinery employees for a moment, Table 5 shows that the number in Detroit increased by 33 percent between 1960 and 1968, and decreased by 20 percent from then until 1972. In Michigan, for the corresponding time periods, there was an increase of 16 percent followed by a 1 percent decrease. In the United States, the number increased by 18 percent between 1960 and 1968, and increased another 2 percent between then and 1972. Even though Detroit figures are not available for the other classifications listed in Table 5, evidence indicates that a comparable pattern exists. These data include employees in nonunion independent firms, unionized independent firms, and captive shops. Most of the decreased employment in the Detroit area prior to 1972 occurred in independent shops.

In the early and mid-1950s, much of the decreased employment arose from the closing of a number of automotive manufacturers.[3] More recently the decline has been exacerbated by the general downturn in the marketplace, but probably more by the tendency of some of the largest customers to expand their captive tooling operations and decrease demand for outside assistance. Total hours worked in Detroit Tooling Association shops dropped from 28.8 million in 1956 to 7.6 million in 1973.

Conclusive evidence for measuring the quantitative growth of captive shops is unavailable. However, information disclosed by the Big Four automakers at the 1969 U.S. House of Representatives Hearings[4] permits an estimate of the expansion of two of the four automotive manufacturers. In-house special toolmaking at one company increased by 50 percent, from a yearly average of $28 million in 1965–66 to an average of $42 million in 1968–69; the other company increased by 95 percent, from a yearly average of $22 million in 1964–65 to $41 million in 1968–69. Additionally, the number of toolmakers employed by one of the Big Four increased from 23,000 to 33,000,[5] approximately 43 percent, between 1958 and 1970. This expansion occurred during a general growth in the demand for tooling, so the impact on the independent shops, though important, was emphasized until the downturns in demand which occurred subsequently. Much of the captive shop growth appears to have occurred in the large body tooling activity.

The extent to which the livelihood of workers in independent tooling companies is tied to automotive needs and policies is indicated by one finding from an earlier University of Michigan investigation:

The term "independent," as applied to those shops not owned by a parent company, needs special interpretation in the Michigan market, where the persistent and specialized demands of one huge industry have supported many tool and die firms. This is especially true of Detroit area firms, 51.5% of which sell over half their output to the automotive industry. If sales to the suppliers of the automotive industry are included, 85.6% of the firms surveyed sell more than half, and *25% all, of their output* to the transportation industry.[6]

It is not surprising that tooling workers intensely follow rumors about automakers' plans to modify a car line or to introduce a completely new one. A paucity of new models means layoffs and toolmaking plant closings, while continual changes, such as those that occurred during the mid-1960s, signal exciting mechanical challenges, full work force utilization, and extended overtime premiums—a magnetic attraction to the many energetic toolmakers. In human terms, the production of tooling for one new car model may employ 3,000 to 4,000 workers for a year and provide all the work 150–200 typical independent companies can handle.

NOTES

1. General Motors has reported that it has produced up to 7 million major pieces from one body die.

2. Changing a major body die typically takes five to thirteen hours.

3. After World War II, nine U.S. automotive manufacturers tooled up their production lines to produce cars; in 1975 there were four. Crosley, Kaiser-Frazer, Packard, Hudson, and Nash all ceased or merged operations during the early and mid-1950s. It is important to observe that these auto manufacturers relied on independent companies for their body tooling needs.

4. Hearings before the Subcommittee on Special Small Business Problems of the Select Committee on Small Business, 91st Congress, 1st Session (1969).

5. It should be noted that some of these toolmakers are performing tooling maintenance, adjustments, and changes, as well as tooling construction.

6. Donald N. Smith, *Technological Change in Michigan's Tool and Die Industry* (Ann Arbor: Institute of Science and Technology, The University of Michigan, 1968), p. 21. (Emphasis added.)

III

THE NATURE OF COSTS AND THEIR CLASSIFICATIONS FOR DECISION MAKING

No matter what the nature of the decision facing the businessman, a basic understanding of the nature of costs and how they behave under differing circumstances is essential to informed decision making. An understanding of how costs and volume are related to profit is essential to profitable business operations because of vigorous competition, the increasing rigidity of costs in automated production, and the pricing decisions that must continually be made in tool and die shop operations. It is also important to understand the nature of costs and how they react to a variety of circumstances in order to comprehend fully the discussions presented in the chapters dealing with the economics of the tool and die industry in general, and the comparability of independent and captive shops in particular. These discussions lead to the ultimate conclusions reached in this study. In this chapter we consider the nature of these relationships and some of the analyses that are made possible by classifying costs according to their tendencies to change or remain constant. In Chapters 4, 6, and 7 we will utilize some of these techniques in considering product pricing in the industry, and the relative economies of independent and captive shops.

The most important distinction among costs for most business decisions involves the classification of costs in terms of their relationship to volume of production or sales, i.e., the ways costs behave as volume changes.

Fixed Costs

Fixed costs are those which stay relatively constant in total dollar amount over given relevant ranges of production activity. Thus, they would be the same at the bottom of the relevant range as they are at the top, or at any volume level between the two. They are the costs of getting ready to do business, including the costs of promoting and influencing future business. For example, when a business organizes and acquires buildings, machinery, and equipment, decisions are made on the basis of the volume of activity which the firm anticipates. Although the amount of the investment will depend on the anticipated volume of business, the initial outlay is necessary before the firm is ready to do business, and the plant and equipment must be maintained regardless of the volume of business at any particular time. Such costs relate more to the passage of time than to the volume of activity within a specified period of time. Once these initial outlays have been made, the investment is "sunk" and the attendant depreciation charges are uncontrollable. The investment cannot, under usual circumstances, be retrieved without substantial loss. Moreover, the acquisition of such items creates the requirement of continuing cash outlays for taxes and insurance. These costs usually continue even if the plant is completely shut down. They can be reduced only over an extended period as capacity is gradually reduced by retirements and acquisitions and replacements are restricted. Thus, a significant portion of fixed costs results from the initial provision of capacity necessary to operate.

Most of the conclusions reached above also apply to the outlays for salaries of operating personnel, although these costs are more controllable than those arising from the acquisition of plant facilities and equipment. A minimum number of operating and office employees must be retained regardless of the volume of output. To this extent these costs are fixed, although only within a known range of volume. As volume increases, a point is reached where additional personnel are necessary. Thus, as volume fluctuates, salaried personnel costs fluctuate, but in sizable discrete steps rather than in small continuing increments. These costs may be treated as fixed costs, although it is recognized that there are levels of volume beyond which they will change. This kind of cost behavior is often referred to as a semifixed pattern.

Expenditures for such things as advertising, personnel training, safety programs, and special research projects do not vary with volume except, perhaps, in the very long run. Although these costs are important and justified, they can be reduced drastically for certain periods if such reduction is necessary for survival. Costs of this type should be considered fixed. In some instances, however, it may be useful to subclassify them as programmed or managed costs to emphasize their discretionary nature.

Fixed costs, therefore, include primarily the costs resulting from expenditures made to provide for anticipated volume and to assure long-run operating benefits. Many of them are "sunk," in the sense that they are inescapable after a major expenditure has been decided on. They largely involve the cost of providing a capability to produce and sell, rather than the costs of continuing operations.

Although the term *fixed* might seem to imply that such costs never change, such an assumption would be incorrect. Few costs are fixed over all volume ranges, and all costs can be altered over an extended enough time period. Thus, management must be concerned with a relevant range of activity and a given time period (usually a year) when it decides whether or not a cost is fixed.

Note again that fixed costs remain relatively constant in dollar total per period. If the accounting system in use includes these costs in the costs of jobs produced, and if the cost per unit is derived by dividing total production costs by the number of units produced, the costs per unit will be high when production volume is low and low when production volume is high. In other words, the average fixed costs per unit decreases as volume increases, and increases as volume decreases. This characteristic of fixed costs is important since it emphasizes the importance of operating at high volume levels. In order to reduce confusion, however, a manager should generally think of fixed costs in terms of a total pool of costs, rather than on a per-unit basis, when setting objectives, controlling costs, or quoting prices.

Variable Costs

Many costs are directly related to production or sales volume and vary in total amount with changes in volume on an approximately proportionate basis. These are variable costs. When volume of production increases 10 percent, variable costs will increase 10 percent, and so on. This means that costs which are truly variable will remain constant per unit, and for most decisions made by the manager, they should be considered in this per-unit sense. Examples are raw materials, direct labor, factory supplies, sales commissions, delivery costs, and electricity used in running automated equipment.

Material costs should vary almost automatically with volume changes if the mix of jobs and the length of production runs remain relatively constant. Many other variable costs, however, have a tendency to creep up per unit as production decreases. Direct labor costs, for example, are normally variable, but restrictions in labor contracts tend to cause efficiency to lessen as production decreases, resulting in an increase in direct labor costs per unit. Also, payments which continue for a time after the worker is laid off, such as Supplemental Unemployment Compensation, tend to move direct labor toward the fixed cost category. In addition, even though, ideally, one can expect that additional personnel will be hired in response to the need for a larger volume of activity and that the payroll will be decreased as volume of operations is reduced, management policy may make labor costs essentially fixed. If the labor force is maintained intact, regardless of volume fluctuations, in order to avoid other costs—such as those of dismissal, rehiring, retraining, and unfavorable unemployment insurance rates—then labor force must be maintained at all times, and a basic fixed portion of labor costs may be expected to exist. Nevertheless, management, in response to changing volume, should be able to keep these costs relatively constant per unit, and thus the variable cost grouping is justified.

Mixed Costs

Some costs fluctuate in small continuing increments as volume changes. These are mixed, or semivariable, costs—i.e., they reflect a mixture of both fixed and variable patterns. Special consideration and attention may be needed to determine the behavior of these costs. Costs of water, electricity, and maintenance are likely to change as production levels change, but they will continue at some low level even if tooling operations are temporarily shut down. For costs such as these it may be useful to set up both a variable portion and a fixed portion and to deal with them in separate segments.

Most managers should be able to classify their costs into fixed and variable categories by simply considering the nature of each cost and observing how it reacts as volume changes. In some instances, however, the plotting of costs of past periods in relation to volume levels of past periods (scatter diagrams) will prove helpful. Figure 3 shows such a procedure.

These scatter diagrams show that material costs are variable and rent (or, depreciation) on buildings is fixed. Factory supplies reflect mixed tendencies. These call for decisions about classification. When a line is fitted approximately to the cost observations, a reasonably accurate *straight* line results. The lines do not intersect at the zero point, however, so some fixed cost tendency is reflected. Factory supply costs, therefore, should be included in both fixed and variable groupings. The estimated cost line for factory supply costs shows a $1,800 fixed-cost element. This amount, possibly adjusted for known changes in conditions, may be considered the fixed portion, which will stay constant at all levels of volume within the relevant range, and the remainder of total supply costs are the variable portion.

The foregoing examples give a very brief and limited explanation of the problem of analyzing the behavior of historical costs. More detailed and sophisticated techniques for handling the problem are discussed in many current publications.

A few warnings

Several problems arise in connection with cost behavior analysis. One relates to the factor used in measuring production volume. In the illustrations presented, direct labor hours were used. The measuring factor should be that element which is most readily available and to which the studied cost is most logically related.

Another problem involves the time segments and the total time interval used. Under usual circumstances, monthly cost patterns for the past

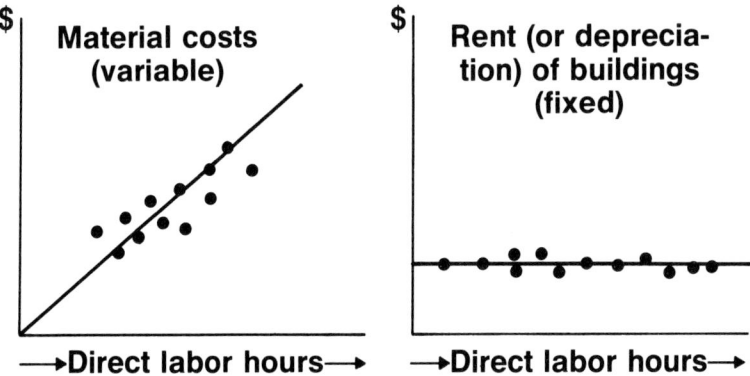

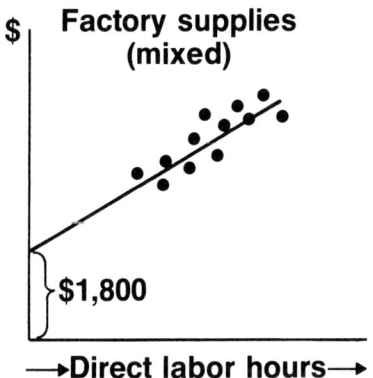

Fig. 3. Scatter diagrams of past periods' costs and volume.

one or two years are appropriate. However, substantial changes in the production processes or in accounting practices may be significant in the choice of the time interval. Therefore, the objective should be to choose periods which are representative of expected future conditions, since such periods provide a cost history that is likely to repeat in the future, and since the anticipated future costs are critical to analysis.

Furthermore, it should be recognized that the patterns of behavior of some recorded costs reflect arbitrary decisions regarding accounting methods. For instance, recording depreciation on a straight-line basis gives the impression that such a cost is fixed (totally unresponsive to volume), whereas accounting for depreciation on the basis of machine usage causes the amount to reflect a variable pattern moving in direct relation to volume. It is the nature of the item, not the way it is calculated, which determines whether it is fixed or variable. Therefore, in accounting for depreciation or other charges where alternative accounting methods are available, the method chosen should be the one which most nearly reflects the expiration of value or the benefit received. If this is done, fixed and variable groupings based on reflected cost patterns will provide accurate measures for decision making. If other methods of accounting are better for income tax or other purposes, appropriate adjustments may be made.

Of course, all costs are variable in the very long run. Plans must be drawn for short periods (yearly), however, and many decisions should reflect the expectations of the short-run behavior of costs. Fixed and variable classifications provide a practical tool in managing a business, even though they are not completely sound as theoretical concepts and even though a high degree of accuracy, in the mathematical sense, cannot be expected.

As discussed later in this study, the classification of costs provides the basis for various analyses. The use of these classifications, once established, usually provides no verification of their accuracy. Faulty classifications may lead to poor decisions and serious consequences. It is thus important that careful attention be given to devising the classifications and making sure they remain accurate.

Cost behavior reflects particular circumstances with regard to the nature of production processes, management policies, accounting practices, and cost incidence. As these factors change, cost behavior patterns may be expected to change. It is therefore essential that cost behavior and cost classifications be reviewed periodically, if not continually, to assure a sufficient degree of accuracy in the analyses which use the classifications.

Cost Analyses Using Fixed and Variable Cost Classifications

The use of cost behavior information for pricing decisions is considered in Chapter 4. We consider here only the basic profit sensitivity analysis or breakeven analysis made possible by behavioral cost classifications. Having examined the relation between costs and volume, we now consider the interrelations of these factors with revenue and profit.

The interrelations may be simplified in a single chart as given in Figure 4. This diagram shows the level of profit or loss that may be expected at any sales level up to $200,000. This amount would include some items of departmental and factory overhead, most of the building and occupancy costs, and selling and administrative expenses. Variable costs are expected to be 60 percent of sales, or $120,000 if sales reach $200,000. The sales volume necessary to break even is $125,000. This may be calculated by dividing the $50,000 fixed costs by 40 percent, the profit contribution rate. At this break-even level, variable costs amount to 60 percent of $125,000, or $75,000, leaving $50,000 to cover exactly the fixed costs. If sales fall short of $125,000, a loss may be expected. A profit should result from sales levels over $125,000. Expected profit may be calculated readily by comparing the profit contribution (40 percent of sales in this case) with the $50,000 fixed costs. Profits may be expected to change by an amount equal to 40 percent (the contribution rate) of any change in sales.

A slightly different way of visualizing these same relationships between costs, volume, and profit is shown in Figure 5. Here fixed costs

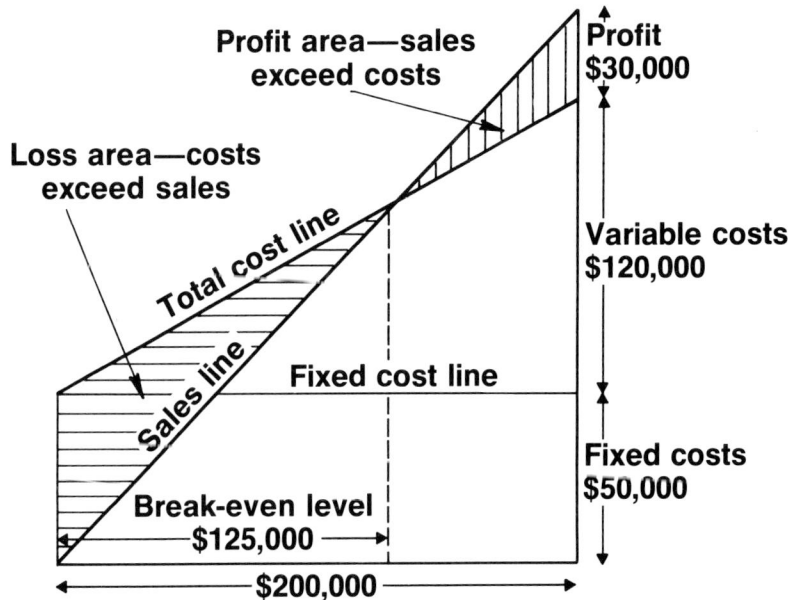

Fig. 4. Cost and volume as interrelated with revenue and profit.

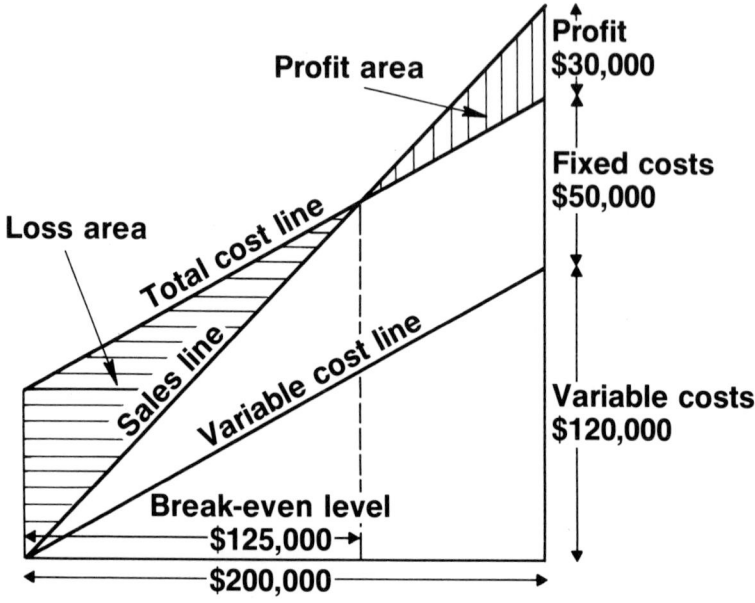

Fig. 5. Alternative view of the relation of costs, volume, and profit.

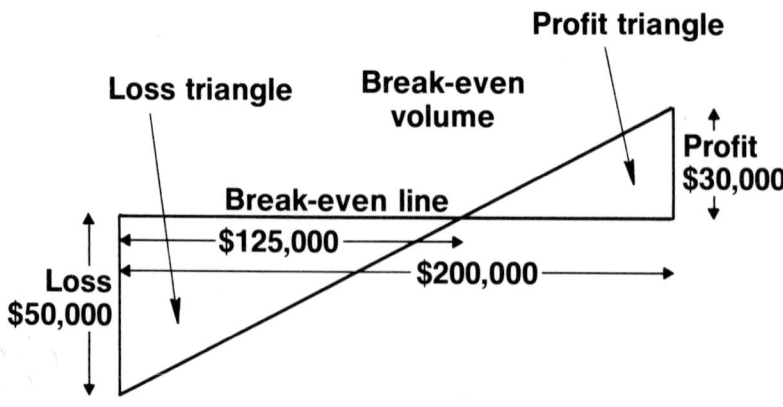

Fig. 6. Simple profit-volume relationship reflecting sales and cost patterns.

are superimposed on variable costs. This more clearly shows the contribution margin (excess of sales over variable costs). At the point where this contribution equals the fixed cost, the operation will break even. This is the level at which the total cost line and the sales line intersect.

These facts are pictured in summary form in Figure 6. Here the fixed and variable costs and the sales lines are not shown. Only the profit and loss line is used. This diagram emphasizes the simple profit-volume relationship, which, of course, reflects the sales and cost patterns.

The break-even analyses already discussed are useful in thinking of overall tool and die operations. Several critical assumptions are involved and must be recognized, however. In addition to assuming the accuracy of cost behavior, we assume a constancy of efficiency, unchanging input and output prices, and a stable mixture of tool jobs. These factors are continually changing, so the analyses must be used with discernment to prevent misinterpretations.

Break-even analyses should be applied at departmental levels or in terms of the particular type of tooling operation involved. Practical considerations will determine the extent to which the analysis may be applied at segmented levels, so long as the increased accuracy gained by these detailed breakdowns and analytical studies is recognized.

Conclusions

Fixed and variable cost classification and related analyses can be very important to the assurance of profitable operations in a tool and die job shop. The tool and die industry, with its large proportion of relatively small operators, is presently not utilizing these analyses nearly to an extent that might improve the effectiveness of the industry. Almost every respondent to the questionnaire indicated his company did not break costs down into fixed and variable components.

Viewing potential profitability in a setting which recognizes the nature of cost behavior is essential to the smallest operator and is well within his reach. The approach is not complicated or difficult to understand. It improves the general manager's ability to think of the total effects of his many important business decisions.

Most of the accounting systems and reporting practices now in use in the tool and die industry do not involve either separations of costs into fixed and variable groups or any recognition of cost behavior patterns, as mentioned above. Accounting practices which may be developed to facilitate reports of a conventional nature and the preparation of tax returns should not be allowed to limit the effectiveness of the accounting or information function in a tooling operation, regardless of size.

The use of fixed and variable cost classification is most important in answering the make-or-buy decision of automotive companies and others who have established captive shops, as later chapters demonstrate.

IV

PRICING POLICIES

Over an extended period of time, each firm, and indeed each industry, must recoup all of the costs incurred in providing a product or service and achieve a rate of return on investment commensurate with the risks inherent in the undertaking, or capital will flow into more profitable opportunities. If a firm is unable to do this, it will go out of business. Likewise, if enough firms in an industry fail to do this, the industry itself will cease to exist.

If a firm is to operate successfully, it must perform many tasks well. One of the most difficult of these, certainly, is the effective pricing of jobs and services. Weaknesses in pricing can cause losses and eventually business failure. Incorrect pricing policies of one or a small group of firms can have far-reaching, detrimental effects on large numbers of firms in a particular locality, especially when the firms are concentrated geographically, as is often the case in the tool and die industry. The statistics reported in Chapter 1 indicate that cutthroat pricing ranks high among the major problems facing individual firms and the industry as a whole. Respondents to the questionnaire felt that this problem arises primarily because (1) a substantial number of managers lack sufficient knowledge of the cost structure of their firms to price effectively, and (2) excess capacity brought about by increased captive shop activity encourages depressed pricing. This is a serious situation. The following statements summarize respondents' answers to the questionnaire with regard to the major problems created in their firm or in the entire industry by poor pricing policies:

1. In many cases plant managers do not know their costs, resulting in cutthroat pricing by many firms.
2. We cannot command a sufficient price on many of our jobs because of fierce competitive pricing.
3. Many customers will get five quotations for a job. Invariably at least one quotation is too low because that manager made errors in estimating the job, or doesn't understand his costs, and he gets the work.

4. We have a lack of work, and resulting low profits, due to improper pricing of competitors.

These statements indicate the severity of the problem as perceived by managers of tool and die shops. It is doubtless true that many firms would find it advantageous to reconsider their pricing policies and make adjustments and improvements where necessary.

If we existed in an atmosphere of pure competition, pricing would not be a problem. The forces of supply and demand would set prices more or less automatically under such conditions. This is true since both buyers and sellers would have full and complete knowledge of market conditions. Buyers would know exactly the worth of the product or service to them and all the prices of competing sellers. Sellers would not only know their costs completely, but also would know those of their competitors. One firm could not establish prices above the competitive price since customers would go to another firm to buy the product or service. If the industry overall tried to sell at prices above the competitive price, the market would not accept the product or service and prices would have to be reduced. If one firm set prices too low, demand for its product would soon exceed its capacity to produce, and prices would rise until demand dropped to the point where sufficient capacity existed to satisfy it. The same would hold true for an industry.

These pure competition assumptions often do not hold true in our economic system, however, so pricing is a critical task if sellers are not to act irrationally. Irrational pricing behavior by an industry, or a firm within that industry, is not good for the customer or the industry. Each tool and die firm must take an active role in establishing rational prices for its products and services. The purpose of this chapter is to provide some guidelines for reaching that objective. It is not possible to provide specific answers which would serve under all circumstances, of course, but the approaches and guidelines established in this chapter should aid in making rational decisions under a variety of circumstances.

A General Philosophy of Pricing

A number of fundamental ideas regarding a general philosophy of pricing must be understood before a firm attempts to establish prices for specific products and services.

A basic consideration for a firm that is establishing a pricing philosophy is to determine its main objective. Although arguments have been presented, pro and con, the basic objective a firm ought to seek is maximization of long-range profits within acceptable moral and required legal boundaries. Without question survival might be of paramount importance under certain conditions. In the short run it also is true that a firm may be more concerned with pricing to enhance the company's image, increasing its share of the market, or growth, as long as profits are satisfactory, even though below the maximum possible. Frequently, however, even policies such as these are really tantamount to pricing for long-range profit maximization. Certainly a firm must achieve a long-run rate of return commensurate with the risks inherent in its operations, or fail. A successful rate of return is more likely to be attained if long-range profit maximization is accepted as the goal.

Frequently things assumed to be true are actually oversimplifications of real-world circumstances. Some believe that a desired profit figure is simply added to costs incurred in producing a product in order to establish its price. In fact, many respondents to the questionnaire indicated this is exactly the way they try to establish a selling price. This would be true only in a monopolistic environment, if at all; certainly it is not the case in the tool and die industry. Even in a monopoly situation, things are not that simple because of joint cost problems. Costs do not control prices charged, although they do frequently influence them.

Others seem to believe that prices have complete control over costs; that sellers take the established price, reduce it by a necessary profit margin, and this establishes the costs of production, i.e., the amounts the sellers are able to pay for materials, labor, and capital. This only approaches being true in a purely competitive situation, and we have established previously that the present environment is not one of pure competition. Prices do not regulate costs, although they do influence them.

Thus, prices and costs influence one another, but usually with neither controlling. In addition, both frequently influence, and are influenced by, volume. When costs increase because of decreasing volume or higher overhead, they often cannot be passed on to the buyer. Selling price decreases, on the other hand, might increase production volume, decreasing the average cost per unit produced and increasing profits. In sum, often there is no single price that can be established. There is, instead, a range of prices within which a sale can be made. Certain costs do establish a minimum below which a price normally should not be established. The upper limit of prices are established by customer demands and the absence of competitive pressure. If competition is severe, as when excess capacity exists in an industry or a segment of an industry, prices may be pushed down to a level nearly equal to out-of-pocket costs, presently a fairly common situation in the tool and die industry. If demand is high for a product or service having no substitutes, price may rise far above the costs incurred in producing and selling the product or service.

These considerations lead to a third consideration in the general philosophy of pricing. The price maker should keep abreast of market conditions generally and the conditions in his industry in particular, especially long-range forecasts regarding both. He should strive to gain whatever knowledge he can of customer desires, the utility of his products and services to the customer under varying circumstances, to estimate and anticipate his competitors' costs, and to know his own firm's costs and how they react under differing circumstances. Perfection in this regard is not possible, of course, but the best price maker is the most informed one. Certainly rational prices cannot be established in a vacuum, i.e., in an environment where knowledge is completely lacking.

As an aside, perhaps, the responsibility of the customer should be mentioned. There is no reason why a customer cannot negotiate and help determine price. There is also no reason why a customer should not take advantage of bargain and distress prices when they exist. Customers should not expect sellers to make such prices continuously available, however, nor would it be in the customers' best interests in the long run to do so. A seller who constantly sells at distress prices would receive a return lower than that commensurate with the risks assumed, or, possibly, the prices would not even cover his costs. In either case, the results would be the same: a gradual liquidation of the business. This liquidation might take some time, but inevitably the supply of the product would be cut off, because capital would be put into other, more profitable ventures. When this occurs, particularly when the service or product is to become a component of a customer's product, the customer may be forced either to drop the manufacture of the particular product or to set up a captive shop to perform that particular service or produce the needed component. Whichever course he takes, his profits are likely to be adversely affected. In other words, the customer, in his zeal to force reductions in price in the short run, may be seriously endangering the long-run supply of a service or product needed to maintain his operations. Certainly we have seen this happen in the tool and die industry, particularly in those firms constructing the large body dies for the automotive industry. The result has been a substantial increase in captive shops since the late 1950s, with the commensurate problems noted earlier in this report.

In setting prices, the price maker generally should anticipate the future prices of his cost components. Past and even present costs are not usually the best cost figures to use in price determinations. This is particularly true in periods of rapidly changing prices for industries whose production processing period is substantial in length, as is the case with certain types of tool and die making. Anticipated replacement values for raw materials, labor, and investments in plant, property, and equipment should be used. Otherwise, capital will erode since higher prices and wider profit margins are needed to generate the funds required to replace these items.

The price maker also must have reasonable knowledge of the cost characteristics of his firm and must use the appropriate concepts of cost. Costs which are most representative from a planning and control viewpoint are not usually the best cost figures to use in pricing. They may be a good starting point in developing such costs, but usually they must be modified and adapted before they are useful in pricing decisions.

A final basic principle is that costs should be broken down for each item in the line and for each type of tool and die making operation, and used to serve as guides in pricing. If this is not done, losses on some types of tooling work may go unnoticed because they are being covered by the high margins on others. A breakdown of costs helps to determine which market emphases should be reduced or dropped and which should be promoted and expanded.

Using Cost Data for Pricing in Independent Shops

Even when we narrow consideration from a general philosophy of pricing to specific consideration of pricing in independent shops in the tool and die industry, relatively broad generalizations still are necessary since the activities of independents in what is known as the tool and die industry cover a large variety of work, are located throughout the United States, and are managed by persons with widely differing attitudes and management philosophies. Nevertheless, the considerations which follow should be helpful in all these diverse circumstances and locations. Because of the special nature of their current and projected situation, special attention will be given, at the end of this segment, to captive shops and to the independents producing large body dies.

We have emphasized long-range profit maximization as a goal. We also have emphasized the difficulties of attaining this goal because of uncertainties (in demand and supply) in the marketplace. A good place to start, nevertheless, is by establishing a target rate of return, on assets employed by the firm or on the owner's investment in his business, on which to base target prices.

Establishing a target rate of return is not a simple process; a number of considerations should be taken into account, such as the return that might be earned on capital invested in ventures of comparable risk, weighted cost of capital, and published normal or average rates of return for firms in the industry. We emphasize that this is a starting point, only. The target rates of return and the target prices based on them are not meant to imply that such returns can be earned on every sale, or even that an average rate of return can be earned over the short run on all sales. They do specify a goal which must be attained (or nearly so) over the long run if the firm is to stay in business. Target rates of return of 20-40 percent before taxes on total assets employed are fairly common in industry under normal economic circumstances. (Certainly these rates fell during the depressed 1973-75 period.)

To establish target prices, consideration also must be given to ratios of investment turnover and profit margin on sales. Suppose, for example, the tool and die shop has a balance sheet as shown (top of page 33). If a 40 percent rate of return on owner's investment before taxes is considered a reasonable target, and if sales for the year are expected to be $620,000, or twice the investment, then a 20 percent margin on sales, or a

Tool & Die Shop Balance Sheet
as of December 31, 19–

Cash	$ 15,000	Payables	$ 70,000
Receivables	22,000	Capital Stock and Retained Earnings	310,000
Inventories	43,000		
Plant & Equipment	300,000		
	$380,000		$380,000

markup of 25 percent on cost, is the target. The relationships and concepts may be shown as

Investment turnover = Sales/Investment
Return on investment = Profit/Investment
Margin on sales = Profit/Sales

or

Margin on sales = Return on investment/Investment turnover.

In this illustration, the target rate of 40 percent is equal to the desired profit of $124,000 divided by the $310,000 investment. The investment turnover is the planned sales of $620,000 divided by the $310,000 investment, which equals two turnovers a year. The required margin on sales is $124,000 divided by the planned sales of $620,000, or 20 percent. The required margin on sales is the required return on investment (40 percent divided by the investment turnover of two), or 20 percent. To convert margin on sales to markup on cost, we divide the margin on sales by the cost (sales less the margin on sales):

20 percent ÷ 80 percent = 25 percent.

Obviously if prices could be set to derive a 20 percent margin on sales for each job or service performed, then the tool and die shop would finish the year by achieving its 40 percent target rate of return before taxes on owners' investment (assuming the figures in the illustration). This would be possible only in rare instances, since many factors other than costs influence prices, as we have discussed. The objective is to price so that the profits on all jobs combined will average the target rate of return. This raises the important question: Which costs are significant to the price maker for individual jobs and services where more volume is needed, competition is severe, and close figuring is essential? Costs should be used to indicate ranges within which prices may be set with known profit results, in addition to providing a basis for calculating target prices. This requires that costs be analyzed to show their fixed and variable natures so that incremental costs can be measured. Assuming such a breakdown has been made, we can look at an example of the determination of a target price as well as the determination of a range of possible prices for a particular job.

In some instances target rates of return may be applied to incremental costs rather than to total cost. This practice may be particularly useful in a shop where a number of different kinds of tool and die work are done and where competition for the different operations varies substantially. Another important advantage is that this method is free from the distortions possible when total job cost figures are used and many overhead cost items are allocated. For example:

Job 121

Material Cost	$10,000
Labor Cost (2,500 hours @ $6/hr)	15,000
Incremental overhead (based on a percentage of labor, in this case 40)	6,000
Total incremental cost	$33,000
Fixed overhead cost (based on a percentage of labor, in this case 60)	9,000
TOTAL COST	$42,000
Target profit (20 percent of sales *or* 25 percent of costs)	10,500
TARGET PRICE	$52,500

For this example, the percentage markup on incremental costs is 31.8 ($10,500/$33,000).

Ideally, management would determine the price needed for each of its operations in such a way that in total they would generate the target rate of return on investment. Even though this can rarely be done, it is a useful objective. Thus, the primary use of costs is to indicate whether management will ever accept a job at a given price and under what conditions certain jobs will be accepted at certain prices, as well as the goal or objective in pricing. (It is to be remembered that target prices will have to be adjusted for competition, special circumstances such as high or low demand, one-shot jobs, and other factors on the demand side of pricing.)

Incremental costs

Incremental costs (immediate cash outlays) to perform a given type of operation must be known, because they usually represent the minimum price that can be charged for the job. Furthermore, the price should be lowered to this point only under certain compelling circumstances. The results of a decision to charge this minimum should also be clearly understood. When a product or service is priced at the incremental cost level, there presumably would be no effect on profits. There also is likely to be no immediate effect on the working capital of the firm. The long-range effects, however, may be very serious. First, the investment in plant, machinery, equipment, and all other items previously acquired, is being eroded. If such pricing policies were continued for an extended period of time, the business would eventually liquidate, since the cash inflow from sales would be inadequate to replace plant and equipment. Second, although prices are easy to lower, they are often difficult to raise. Once prices have been established at an incremental cost level, customers may expect them to continue and may vigorously oppose increases. Third, cutting prices to this level would force competitors to do likewise and would precipitate a downward spiral which might be very difficult to counteract.

Yet there are cases when a firm might logically sell at a price approximating incremental costs. When an entire industry is in a serious economic slump and incremental cost is as high as the market is willing to go to acquire the product or service, firms might accept business at this price level in order to keep their work force, management personnel, and supply lines intact. Equipment and machinery often continue to depreciate when not in use (and some depreciate more when not in use), so the firm might be better off operating under these conditions than closing down.

Even if a firm never sells at this low level, there is another important reason for having information on the incremental costs for each type of operation. Any price above this level contributes to the fixed overhead. Knowledge of profit contribution is necessary if the firm is to determine the mix of tool and die operations that will lead to maximum profits. Management is then able to emphasize the type of tool and die work which makes the greatest contribution toward profit, and to reduce or drop certain types of work when the market is unable or unwilling to pay a price which makes a reasonable contribution. The relative weights of material, labor, factory overhead, distribution, and selling expenses should be considered, because they vary with each type of operation and also over time. Furthermore, unless a more nearly complete cost per type of job or operation is known, it is impossible to know if one operation is subsidizing another or why the price of a particular operation is not competitive. Full cost per type of operation is important to permit management to see the consequences of a pricing decision. This information allows management to make a more informed decision about accepting or rejecting business at a given price.

Here, again, certain conditions may influence management to set a price between incremental costs and full costs, though this is not an ideal practice. If there is idle plant capacity and the job is a one-time order requiring only a short time to complete, management might well decide to take it at a price higher than incremental costs but less than full costs. Every dollar above incremental costs contributes to fixed overhead (and hence profits) and is a gain over what would be obtained if the capacity were allowed to remain idle. A word of caution is necessary. Management should not, as a general rule, accept on this basis jobs which are going to be repeat jobs nor should they accept on this basis jobs which will extend over a considerable period. To accept repeat jobs on this basis (meaning jobs that are handled repeatedly into the foreseeable future) is to establish a price the customer will continue to expect in the future. Jobs accepted on this basis of long duration reduce the flexibility of the firm and make it difficult, if not impossible, to take advantage of high-priced jobs when they are available.

Of course, the objective of management is to sell at a price which will result in achieving the target rate of return on investment (or higher). Since this is the goal, target price information similar to that which we have been discussing should be constantly in front of the price maker so that he can see how close he is coming to the goal. A target price should not prevent his accepting jobs where the market has set a higher price, of course, but this information, along with information on full costs and incremental costs, makes price setting (or the decision to accept business at a price set by the market) more scientific, and it makes the results of the decision clear.

When the economy is running at a high level, the management of a successful firm is tempted to play down or be unconcerned with the pricing problem. Yet the concern for proper use of pricing techniques should be almost as great under these conditions as it is for the management of an unprofitable firm. When a firm is operating at near full-capacity levels, it must exercise great care to ensure a mix of tool and die operations that will maximize profits. The jobs which are then accepted should be those entailing operations which are likely to yield the highest contribution to profit. Otherwise plant capacity will be used up on low-yield jobs, and even though the operations might still be profitable, they would not be as profitable as they ought to be. In other words, a ranking scheme needs to be devised in order to assume maximization of profits.

Moreover, there are bound to be low levels of economic activity during which companies will operate at considerably less than full capacity. (Many tool and die companies are in this position at present.) Cost information and target prices as described above will help such firms to hold profits at the highest possible level (or losses to the minimum possible level). During slack economic periods the question is often not which firms make the greatest profits, but rather which firms suffer the least losses. It is the latter which survive the slump and reestablish their profit pattern when conditions change.

But there are other benefits as well. Pricing by the procedures discussed in this chapter may help competitors decide who has the competitive advantage in various types of tool and die work. In other words, the experience of a particular firm, the type and nature of equipment used, the quality and size of the work force, and the location of the plant may enable that firm to perform a given type of work much more efficiently and economically than other firms in the area. Another firm may be more efficient in another type of work, and so on. In certain instances this may indicate that specialization will result in greater profits, or that two or more tool and die makers with strengths to compensate for each other's weaknesses might merge operations (not necessarily physically) to their mutual advantage, or that a referral system should be established whereby work of a given type is referred to one or a few firms which are the most efficient and economical in that particular operation. Since there appears to be excess capacity in the industry in general, and in the Detroit area in particular, it is essential that each firm perform the operations at which it is most efficient. An allocation of demand among firms according to individual efficiencies is needed. The suggestions in this chapter should aid in accomplishing this.

Since tool and die making is a service function where total volume is fixed by the needs of the industries served, changes of prices within reasonable bounds are not likely to affect total volume of the industry substantially. Cutthroat pricing is therefore not likely to materially increase total volume in the tool and die industry. More likely, it only will switch volume back and forth between individual firms with all the short-run benefits accruing to the advantage of the customers.

Pricing by Independent Shops—Conclusions

One of the main purposes of this chapter has been to outline the major considerations in price setting, to indicate how these might be implemented, and to show the results which are likely to follow the applications of various principles. Once prices have been established, they can be converted into sales price per man hour, per machine hour, or any other base that a particular management may wish. We have attempted mainly to point out the considerations involved in arriving at prices, regardless of how they might eventually be stated.

Although a few firms are doing some of the things recommended here, questionnaires and field interviews indicate a need for improved practices. For example, a vast majority of the respondents to the questionnaire answered that they did not make any distinction between fixed and variable expenses for pricing purposes, and many did not appear to be well-informed about the proper use of such information. Almost all respondents considered historical costs only in setting prices. Few seemed to make any estimates of material, labor, and overhead costs for the coming year for use in establishing prices. Many had little idea of the success or failure of their pricing decisions until they received monthly or quarterly profit and loss statements.

The tool and die industry can benefit from use of more scientific methods of establishing prices which could lead to the elimination of cutthroat pricing, a greater degree of specialization where warranted, and closer beneficial relationships between competitors, such as referrals of business. Achievement of these goals not only can aid in the maximization of long-run profits to the firms involved in the effort, but also can help to improve the industry as a whole.

Independent shops making large body dies for automotive companies

Because of the special circumstances facing shops making large body dies for the automotive companies, a few comments concerning these shops are in order.

Such shops are plagued with substantial excess capacity; this was true even before the slump in the automotive industry of the past two or so years. Even if and when the automotive industry recovers, the situation is not likely to improve for this part of the tool and die industry *unless* either the automobile manufacturers decide to place substantially more work outside their captive shops, or the government takes some action to alleviate the situation. Both of these possibilities appear remote, particularly considering the excess capacity presently existing in the captive shops and the lengthening of auto model style cycles from approximately three to about five years. Many large independent die shops have been surviving primarily on the peak period of the cycle when the captive shops have insufficient capacity and are forced to send work outside. Such peaks, if they arise at all in the future, are likely to be smaller, fewer, and farther between.

Consequently, unless some unforeseen, drastic change takes place in the near future, more and more such shops will have to go out of business until capacity is reduced in total to the point where those remaining have sufficient orders to stay in business profitably. No one knows for certain exactly when that point will be reached, but at this writing some additional reductions in capacity seem highly likely. In a broad sense, two results from this trend are likely: (1) total effective competition will be reduced even further, since more and more tooling will be produced in-house, and will probably result in higher prices to consumers, and (2) total capacity for producing tooling, even if not reduced (and some reduction is likely), will at least be centered in fewer locations. Such concentration, and reduction, has national defense overtones which need consideration beyond the possibilities of this study.[1]

In any case, under these circumstances of excess capacity, and considering the dynamic competitive spirit possessed by independent tool and die makers, cutthroat competition is likely to be commonplace. Survival is the first principle of business, so this is understandable, but it might not be logical. Although the principles stated in this chapter will be helpful in these circumstances, a few other points should be considered by the management of these shops.

If a given shop's cash flow is not at least equal to what might be earned on the investment in a savings account at the bank, and if this situation has existed in that shop for a period of time with little relief expected in the future, the correct decision might well be to go out of business. (Cash flow can be quickly, and roughly, calculated by adding depreciation and amortization of plant, machinery, and equipment to net income). Since there is an active used-equipment market for the large machinery and equipment utilized in these shops, the manager should calculate what he could earn on the funds received from the capital equipment and land if he disposed of them and invested the funds in a savings account at approximately 8 percent. If the cash flow from his business is less than this return on a virtually riskless investment, consideration certainly should be given to closing up shop. In a sense, his business is worth more dead than alive; i.e., it has greater liquidation value than it has as a going concern. Note that we have been discussing a very minimum rate of return to this point. As a matter of fact, the cash flow would need to be considerably higher than such minimums to make continued operations advisable because of the high risks involved in tool and die making. Certainly personal attitudes and feelings are important in making the final decision, but this opportunity cost of staying in business should not be overlooked. The longer one stays in business under such conditions, the more he loses in the sense of the higher rates he could be earning in other investment opportunities.

The decline in profit levels of companies producing large body dies intensified at a time when certain automotive manufacturers were advising their suppliers to acquire highly expensive numerical control (NC) die-making machinery in order to be compatible with emerging computer-aided design and manufacturing methods. The capital required for this costly conversion was totally unprecedented for even the largest and most financially stable independent companies. In the 1967 University of Michigan report, it was shown that the purchase of only one numerical control die mill—at 1965 pre-inflationary price levels—was several times the total invested capital of a typical body die manufacturer.[2] Moreover, few companies converting to numerical procedures could do so with less than two or three NC mills—making such purchases virtually beyond their financial capacity. It may be an omen for the future that, of the few independent body tooling companies with the courage to negotiate the large loans required to buy expensive die mills, virtually all of them subsequently experienced cash flow emergencies during business downturns and closed operations.

Nevertheless, a great disparity continues to widen between the independents on the one hand—strapped by their falling profit levels and dissipating capital resources—and the Big Four on the other—with their enlarging appetite for tooling to be produced by methods compatible with computer-aided design technology. General Motors continues, for example, to make impressive advances in its computer technology which generates mathematical product definitions rather than physical models for machining die components. Without the machining models, the copy milling machines that are used almost exclusively by the independent companies are useless.

General Comments on the Pricing of Tooling in Captive Shops

Although much of the data supporting this discussion on captive shops will be found in Chapters 6 and 7, a few general comments are appropriate here. As a general economic proposition, most of the principles discussed regarding pricing policies are equally applicable to firms having captive tooling shops. Except in rare instances, captive shops do not produce for and sell to customers outside the firm owning them. Consequently, their problem is one of interdivisional pricing, i.e., what price is appropriate to charge to the divisions of the firm for which they produce tooling. As stated, from an economic viewpoint, essentially the same considerations are appropriate in this setting as those which would apply if the captive shops were selling to customers outside the firm (assuming the goal is to minimize costs).

However, interdivisional price setting can be, and is, used to accomplish goals other than production of the final product at the lowest possible cost. One such objective is goal congruence, i.e., setting a price which encourages the managers of the divisions involved to make decisions which are in accord with overall firm policies, even when they, as division managers, might not feel such decisions are in their own best interests. One such policy (stated in Appendix B) is to maintain stability in the work schedules of the skilled tradesmen in order to avoid union problems. To do this the interdivisional price for tooling must be established at or near the price per hour that independent shops would charge for making the tooling. Such a price would encourage the divisions using the tooling to buy from those producing it rather than going to the outside, at least up to the point necessary to keep all present tool and die employees working full schedules. Beyond that point, work would be placed on the outside. If a higher price than this were established for interdivisional pricing, the division using the tooling would buy from the outside and violate the company policy. (It might be pointed out, again, that union contracts with automotive companies having captive shops presently require union approval before work can be let to outside shops). One of the Big Four is presently using a $16.85 per hour rate to transfer tooling between divisions for the 1976 model tooling program. It will be raised to $17.85 per hour for the 1977 model. These rates approximate what independent shops are presently quoting per hour for tooling work.

Note that such rates do not necessarily bear any relationship at all to the actual costs per hour for the captive shops. Chapters 6 and 7 indicate that actual costs may be running from as low as $25 to as high as $50 or more per hour. As pointed out above, if such rates were charged interdivisionally, most work (except for repair, maintenance, tryout, etc.)

would go to the outside (ignoring union contract restrictions). With such high costs, and using a transfer pricing rate of about $17 per hour, undoubtedly a loss situation is reflected for the captive shops. However, if company policy is to build inside, the performance of the captive shop manager can be judged by other means, perhaps by using a higher per hour rate than that for interdivisional pricing purposes.

Does this necessarily mean that *all* work should be sent outside? Not at all. Some basic tooling capacity is needed at the automotive plants for essential tryout and maintenance work, and to assure security for new products. And, certainly, some capacity is needed to assure that the needs of the automotive companies are met on a timely basis, since independents may not always have capacity sufficient to do all required tooling work demanded on the final schedule. It does appear, however, that some retrenchment, some swinging of the pendulum in the other direction, would be beneficial to all concerned. Being in an oligopolistic position, automotive companies, within certain constraints, have been able to pass most cost increases onto the consumer since the demand for automobiles, within limits, has been relatively inelastic. Although certainly not the only factor decreasing the demand for automobiles over the last two years, increasing costs did push the price beyond the limit acceptable to buyers, and demand dropped drastically. Perhaps buyers will become acclimated to the increased prices, and the cycle can begin again, on a higher level. Perhaps not. The point here is that total price to the consumer probably is lower because automotive firms have the basic tooling capacity discussed above. On the other hand, costs would be lower and, therefore, prices charged to consumers undoubtedly would be lower, if more tooling beyond those basic requirements were let to the outside.

In sum, unless substantial additional work is placed on the outside by automotive companies, most, perhaps all, of the present large body die shops are likely to go out of business, taking with them the capacity the automotive companies will need if and when demand creates the peaks previously enjoyed. Such a situation might create serious competitive disadvantages for American Motors (which does not have captive tooling shops), and even for Chrysler, if their present reduction of in-house production of tooling continues. Finally, the data in Chapters 6 and 7 indicate that if more tooling work were done on the outside, overall costs could be reduced, and hence, the prices charged for automobiles.

NOTES

1. The erosion of this important segment has serious implications for the country's ability to produce major tools for any future national emergency. Significantly, several of the large machines auctioned off during the liquidation of independent companies were

shipped out of the country. Some were observed by manufacturing personnel of the Big Four during subsequent visits to other companies in foreign countries.

2. Donald N. Smith, *Technological Change in Michigan's Tool and Die Industry* (Ann Arbor: Institute of Science and Technology, The University of Michigan, 1968), p. 82.

V

MARKET CHARACTERISTICS AND PROSPECTS OF THE TOOL AND DIE INDUSTRY

Introduction

The unique and varying demands of the marketplace have forged the structure and nature of the tool and die industry more than any other single factor. The output of automotive toolmaking companies is directed primarily to the Big Four—American Motors, Chrysler, Ford, and General Motors. At any given time a thousand or more tooling sellers may be attempting to market their products and services to only a few huge buyers.

This situation approaches the economic structure known as a monopsony. Almost the converse of the more familiar term *monopoly* in which there are few sellers and many buyers, *monopsony* describes a condition of many sellers and few buyers. Here markets are so concentrated that a few firms may exert sufficient market power to override competitive forces in determining conditions of sale.

Market Size

The exact size of the market for automotive tools and dies is obscured by many factors. Quantitative market measurements therefore must be based on composite data from related sectors, some of which are also estimates. The estimates derived below have been checked against many sources. They were also tested through alternate computational procedures. Consequently, we feel they are reasonable.

A chief barrier in estimating the size of the automotive tooling market is the comprehensiveness of the term *tools and dies*. It may include not only dies, fixtures, gages, and molds, but sometimes also related products such as specialized machinery and attachments. Because some companies producing in this product range can be legitimately classified as tool and die, machine tool, or cutting tool companies, data on their output may be included by the federal government in the sales statistics of any of these Standard Industrial Classifications. As a result, estimates derived in a special analysis such as this one cannot be checked readily against government data for either accuracy or completeness. An additional definitional problem arises because the auto manufacturers may use still other criteria for assigning expenses to the special tool and die category. These decisions are often determined by federal income tax considerations and state or local property tax regulations.

Market measurement problems also arise because of the Big Four's reluctance to publicize detailed plans, activities, and expenses for model changes. A premature presentation of this information would allow competitors to extrapolate from past experiences and thus identify future styling changes sooner than is desirable from the company's competitive viewpoint. Despite measurement and estimation barriers, information is available to indicate the general magnitude of the expenditures by major buyers of automotive tooling. Total expenditures for all types of special tools and dies are published in the annual reports of the Big Four, except for American Motors. Amortization of AMC's tooling expenses are published, however, so that reasonably good approximations of their expenditures are possible.

As can be seen in Figure 7, tooling expenditures for the Big Four have grown from about $800 million in 1962 to $1.8 billion in 1974.[1] Besides the increasing market size, two other major relationships are apparent: the relatively large proportion that GM represents of the total (between 50 and 60 percent), and the substantial variations in expenditures—e.g., 1969 expenditures were about 20 percent greater than 1968's; 1970's were 35 percent less than 1969's; and 1971's were 20 percent larger than 1970's.

Composite data in the figure mask the extreme variability of tooling specialties. The demand for body tooling is particularly variable. The extent of this variability is evident in Figure 8 showing the yearly changes in volume for material cast for body tooling by one of the Big Four. Since a typical toolmaking company is heavily committed to one of the Big Four's special needs, its sales rise and fall with the particular needs of the primary customer. Tooling expenditures by AMC, Ford, Chrysler, or General Motors normally are even more variable than their composite data.

Beyond these published totals for tooling expenditures, other fragmentary information is available to provide useful hints on the nature of and the market for automotive tools and dies, and body-related tooling. Some of the relevant information follows:

1. There are, for example, 10,800 dies and fabricating tools, costing an estimated $500 million, employed jointly in the manufacture of Chevrolet car bodies and the bodies of other GM car divisions.[2]
2. The annual model change does not mean a complete redesign of all of a firm's car lines, or even of any one of them for that matter,

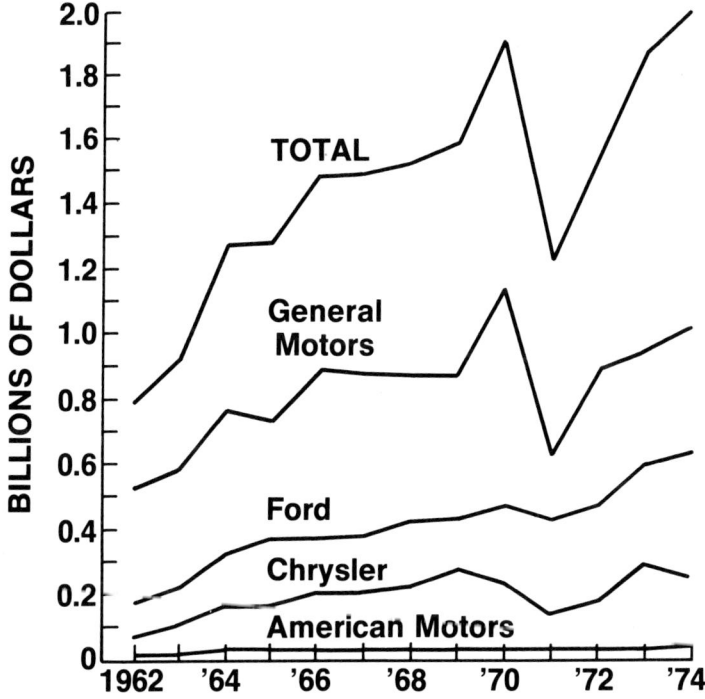

Fig. 7. Annual expenditures for special tools and dies for the Big Four auto makers. (Source: Annual reports from these companies.)

46 / *The Tool and Die Industry*

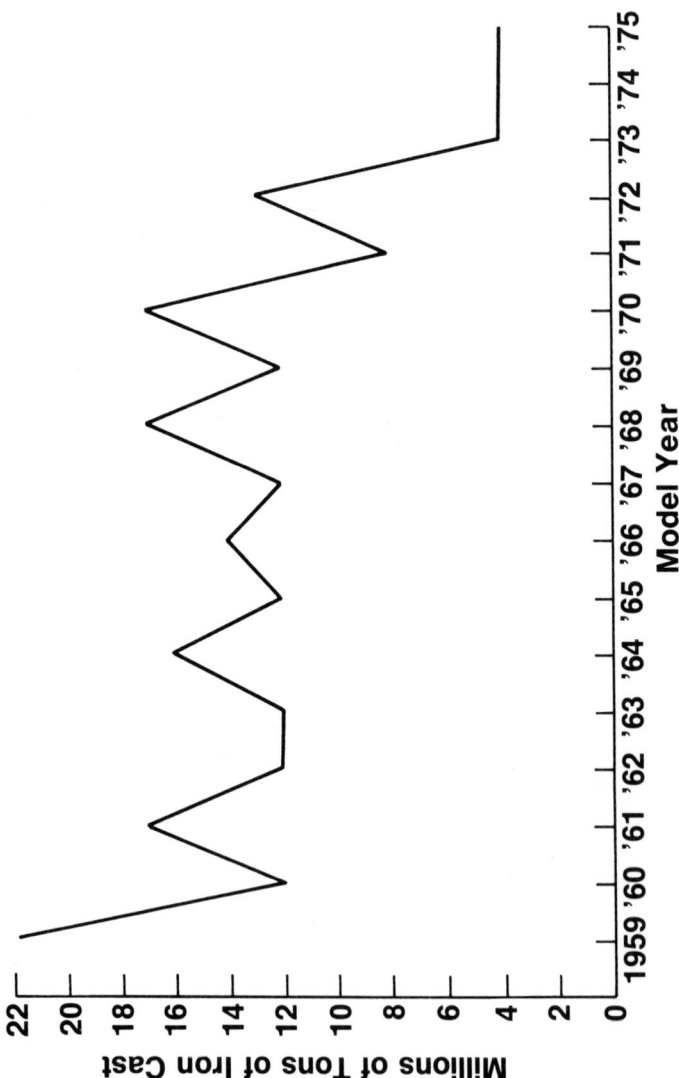

Fig. 8. Variations in model tooling as measured by amount of iron cast for body dies for one of the Big Four auto manufacturers.

every year. It has been General Motors' practice in the past to provide its higher priced car lines with completely new, or at least substantially new, body designs every fourth or fifth year. During intervening years, the appearance of a given car line might be enhanced by major or minor body sheet metal changes. The cycle for General Motors' smaller cars has been longer. The Chevy II (now called the Nova) was introduced in the 1962 model year, received a completely new body in the 1968 model year, and is scheduled for major sheet metal changes on 1975 models. While there were no major design changes on the Nova and other GM compact cars during the intervening years, substantial improvements were made in some components.[3]

3. At hearings before the congressional Subcommittee on Special Small Business Problems in 1969 Frank O. Riley, Vice President, General Motors Manufacturing Staff, stated:

> I would again like to emphasize...that we are discussing...die construction for outer and inner panels and other sheet metal parts for General Motors passenger cars. This represents approximately 75 to 80 percent of our average annual die construction needs for all purposes."[4]

4. An impressive and comprehensive analysis of automotive body tooling costs has been made by John S. McGee under a grant from the Ford Foundation. In his findings McGee comments:

> One European producer estimates that an "all new body" for one smallish model would, in the UK, cost $16 million for skin dies, plus another $4 million for subassembly and welding tools, etc. For a "major re-skin" of an existing model, assuming that the old cowl, for example, could be kept, the die costs would be $12 million to $14 million. According to them, costs would be significantly higher in some parts of the continent and in the U.S., partly because the rate and volume of production would be expected to be higher. According to another producer, for one smallish "new car" in sedan and station wagon versions, in which the mechanical units (power-train, etc.) were mainly carried over, sheet metal dies cost about $21¼ million.
>
> These figures are lower than a third set. According to the latter, to produce one model of a small sedan, around $46 million investment is required for all body dies, plus tools and fixtures for body assembly.
>
> Such figures will, of course, vary somewhat with the size and type of vehicles concerned, and considerably with different production programs planned. But it is fair to infer that body dies for a smallish sedan produced at medium rates and volume now (1974) cost at least $16–$18 million with another $4 million to $6 million necessary to weld and assemble the panels and parts into a body shell. For high rates and volumes these totals rise to around $35 million and $11 million, respectively; figures which nevertheless yield lower costs per unit if the large outputs they permit are actually achieved.[5]

48 / *The Tool and Die Industry*

The evidence obtained during this investigation aligns more closely with McGee's latter figures. For example, it was reported that the tooling for AMC's new 1975 car, the Pacer, cost between $50–60 million.

In summary, it is estimated that the ordering of new body tooling for a new car of American sizes and volumes, at 1975 expense levels, will produce about $35–50 million of business in body dies and molds and another $5–15 million in associated tooling.

Additional market estimation problems arise because not all of the Big Four's tooling expenditures are for automotive purposes, nor do they all occur in the United States. The published financial data cover the parent company and consolidated subsidiaries around the world. Of even greater importance to answering the questions of this study is the lack of available information indicating how the Big Four's published expenditures are divided between the procurement of tooling from outside sources and the construction of in-house facilities. Thus, although the Big Four's expenditures may have totaled $1.8 billion for 1973, some of this was retained in-house, some went for tooling appliances, marine or other nonautomotive products, and some proportion was not available to the U.S. tool and die industry because it was produced by foreign companies. According to information presented at hearings on the problems of the tool and die industry conducted by the U.S. House of Representatives in 1969,[6] the amount of tooling supplied by foreign sources is not great, however. Also, tooling expenditures by the Big Four for nonautomotive operations are small in comparison to automotive tooling needs.

When this information is evaluated, integrated, and compared to other facts and relationships, we conclude that well over four-fifths of the total Big Four's tooling expenditures are automotive-related, and that less than 5 percent of this automotive tooling is procured from non-U.S. tooling sources. We estimate that, when reduced to quantitative terms, approximately 80–90 percent of the total special tooling expenditures shown in Figure 7 is automotive-related and procured in the U.S. The importance of this market segment is indicated by its estimated size; based on the figure of 85 percent of the total being U.S. automotive tooling, the market was around $1 billion in 1965, and $1.5 billion in 1970 and 1973.

Additionally, these data cover only Big Four tooling expenditures. Their component suppliers also require tooling for production lines. This segment, however, is considerably more fragmented; therefore its magnitude is difficult to estimate within the constraints of this investigation. The best available evidence suggests that these suppliers add another 10 percent to the total automotive tooling market, and about 5–7 percent to the body tooling segment.

This, then, is the segment of the market that most of the toolmaking companies in this study group pursue. A strong market exists when several lines of cars are retooled or new models are being readied.

Conversely, tight markets prevailed during the late 1950s, and even more so during the early days of the auto industry when car styles, such as the Ford Model T, were not changed substantially for nearly twenty years.

While the total market has been growing, the body tooling segment has been declining since the late 1960s. In 1974 and 1975 the demand for body tooling declined to post-World War II lows. This decline resulted not only because of the recession, but also was significantly attributable to the lengthening of the body style change cycle from approximately three years to five or six years. Precipitated by many factors, the stretchout resulted mainly from the need to utilize much of the available tooling budget for implementing federally mandated safety and emission standards. The lengthened cycle also resulted from smaller cars with lower profit margins taking an ever greater share of the market. Lower margins generate less cash for styling changes.

Reductions in body tooling procurements were also caused by the increasing use of plastic automotive components.[7] Plastic parts reduce or eliminate the need for several types of sheet metal dies, such as piercing, blanking, and forming, along with the assorted checking, fabrication, and welding fixtures used in conjunction with component manufacture and assembly. An important attribute of a molded plastic part is that it often contains the equivalent of several sheet metal components that previously were assembled together. While mold-making companies gain some new business, the resulting loss in business to traditional body tooling shops can be three to four times the gain.

Market outlook

At present (mid-1975) there are strong indications that demand for automotive body tooling will not soon return to the relatively high levels achieved in the early 1950s and the mid-1960s. While many tooling orders are expected from the changes in reducing the car's body size to help improve fuel economy, much of the available capital for tooling changes will be consumed in retooling drive-train and other nonbody components. Moreover, once weight and size reductions are achieved, the interval between subsequent styling changes appears likely to stretch even further.

Impact of Captive Shops on Tool and Die Markets

Make-or-buy decisions of the Big Four obviously have a major impact on the tooling market. The criteria utilized in these decisions by the American automakers were explored in 1969 by the U.S. House of Representatives Subcommittee which considered the problems facing the tool and die industry. Because all of the Big Four automakers occupy somewhat distinct competitive positions, their make-or-buy practices vary. The practices reported to the committee are summarized in Appendix B.

Exactly how these make-or-buy tooling policies, operating in tandem with the prevailing market forces, have impacted the interrelated volume levels of the captive and independent shops is somewhat obscured by many factors, but there is no questioning the fact that the make decision has critically affected independent companies. What is most obvious is that the companies' policies have changed markedly since the early 1950s, except for American Motors'. There has been a general tendency during this period to transfer the make-or-buy decision and implementation power from personnel at individual operating plants up to a centralized level. Independent company owners indicate that whereas they previously dealt with thirty to forty buyers, they now deal with only two or three at each of the Big Four automakers. One effect of this consolidation has been to lessen the power of the mechanically oriented plant personnel and to enlarge the role of purchasing department personnel. Following from this, less tooling is now procured on technical merits, and more is procured on a purely cost-competition basis. As a result, there appears to be less concern in the strategic assignment of work to assure that companies needed in the future will have adequate work to carry them over during hard times. These conditions also open the door for foreign tooling sources who, because of relatively low-cost labor and direct or indirect subsidies from their governments, are able to enhance exports considerably.

Despite the testimony summarized in Appendix B, which appears to de-emphasize the importance of cost as a criterion in make-or-buy tooling decisions, the shift in the locus of these decisions to the purchasing department has placed a constantly greater emphasis on competing prices of independent companies. The independent tool companies reported that this change in policy in part has accounted for the drastically reduced profit margins in the industry. In some of the buying companies, the change in emphasis reached such a point that they requested independent companies to submit balance sheets and some profit information for yearly review by the purchasing departments. Although this information was required in order to determine whether or not the independent shop was strong enough financially to see a sizable order through to satisfactory completion, undoubtedly the information supplied increased the pressure for still lower pricing on the part of the independents.

While the specific effects of these procurement policy changes on make-or-buy decisions are somewhat blurred by a crosscurrent of several other factors, it is clear that the competitive importance of protecting styling secrets did motivate automakers to make outerskin body dies in the captive shops, at least for a time. This drive to protect styling changes was especially strong from the early 1950s to the late 1960s, when the American car-buying public seemed to emphasize the styling factor in its buying decisions. Although gradually decreasing in importance, as indicated by one automaker's 1971 decision to have several body dies built by a competitor's captive shop, secrecy, together with the desire to have a larger pool of toolmakers readily available, still appears to have weighed heavily in the critical decisions of the expansionary 1960s that led to larger toolrooms for the several new stamping plants built during that interval. These and other related milestones affecting the comparative volume of captive and independent shops are summarized in Table 6.

As indicated in the General Motors Corporation testimony presented at the 1969 hearings, a factor that also influenced the corporation to add to its captive toolmaking capacity was its inability during peak-demand years to obtain sufficient assistance from independent shops. Such a condition led to the construction of perhaps the largest, and certainly the most technologically advanced, toolmaking plant in the world. According to testimony at the hearings:

> ...the decision was made in the fall of 1966 and Chevrolet experienced severe problems in getting started in the model year that fall, primarily due to late delivery of some of the tools and dies. There had been in the Flint area five shops that were no longer available to the Chevrolet division for services they previously performed for them. These shops were taken over by somebody else or went out of business, but not from a lack of work. There was also a shortage of shops that produce large dies at that time.[8]

General Motors subsequently submitted to the committee a detailed statement delineating the basis of its conclusions on outside capacity. Part of that statement follows:

> As our tooling requirements increased, we had to turn to the outside shops to provide an additional tooling capacity. However, we would like to point out three significant factors which indicate that this has not been a suitable arrangement.
> The first has been the gradual decrease in the number of large, independent die shops. During the past five years alone, the Flint plants have lost the services of such shops as Buell, Deluxe, Jeffrey, Koestlin and Mercury.
> A second factor has been the lack of available sources for placement of dies ten tons or over. Dies in this category have increased tremendously during the past few years. This increase, in part, is the result of our effort to maintain the basic car cost. One way this has been accomplished is by combining die operations to reduce piece cost. These large progressive (multi-operation) dies, however, can be built by only 23 percent of the Detroit Area die shops.[9]

Periodic shortages of capacity appear to be a direct consequence of at least two conditions. First, automakers have been delayed, at times, by some independent companies because of technical difficulties, strikes, or, in isolated cases, poor performance (usually when an uninitiated shop misestimated the problem and underpriced the cost of successfully

Table 6
KEY MILESTONES IN TWO DECADES OF EXPANSION OF CAPTIVE SHOP TOOLMAKING

1953	Chrysler buys Briggs Manufacturing Co.
1954	Ford opens Cleveland Stamping Plant Tool Room
1956	Ford opens Chicago Stamping Plant Tool Room
1957	Chrysler opens Twinsburg Stamping Plant Tool Room
1964	GM installs numerical control equipment
1965	Ford opens Woodhaven Stamping Plant Tool Room
1965	Chrysler opens Sterling Stamping Plant Tool Room
1965	Chrysler buys Deluxe Tool and Die Shop
1966	GM opens Kalamazoo Stamping Plant Tool Room
1966	GM decides to build large captive tool room for Chevrolet
1967	Ford buys numerical control equipment
1968	Chrysler buys numerical control equipment
1973	Union wins right to strike over subcontracting of tools and dies
1973	GM closes Plant 23 tool and die operation
1975	Chrysler closes its Warren Tool and Die Plant

building large tooling). The second factor is more fundamental and impinges on the entire industry that supplies automotive tooling. At the base of this factor is the "feast or famine" nature of the automotive body tooling business, and the automakers' (and unions') goal of having as little fluctuation as possible in their in-house workload. Such policies have the effect of placing the responsibility of handling the underload, as well as the overload, on the independent shops. In other words, automotive companies wanted the independent shops available during peak needs for tooling when their in-house capacity was insufficient to handle the demand. This made it possible for them to avoid investing in a total capacity equal to peak needs, thereby avoiding even larger amounts of fidle capacity during much of the year with its attendant costs. When demand was less than could be satisfied in-house, the independents carried the burden of idle capacity.

It is obvious, or should be, that such a policy is self-defeating. Unless such peak needs occur often enough for the independents to make a rate of return on investment commensurate with the risks assumed, they must liquidate. This in turn, requires that captive shops be expanded to meet peak needs. The fact that significant numbers of independent body tooling shops have gone out of business in the past decade indicates

Market Characteristics and Prospects / 53

these peak demands have not occurred often enough, and it appears they are likely to occur even less often in the foreseeable future.

In the past, when the outlook was favorable for a prolonged period of tooling demand, new captive and independent plants were formed, and existing companies expanded their operations to handle the so-called feast conditions. As the so-called famine period of the cycle hit, the independents retrenched and used whatever capital they had as a buttress against recurring losses. If the famine was short, most shops survived to be able to help the automakers during the next feast period. When the famine lingered, the mortality rate was higher, and fewer companies awaited the automakers' calls. The interrelationship of this cyclic phenomenon and the resulting expansionary trends of captive tool rooms are graphically shown in Figure 9.

If unaltered, the consequences of these trends is that eventually all the toolmaking capacity for large body panels may exist only in captive

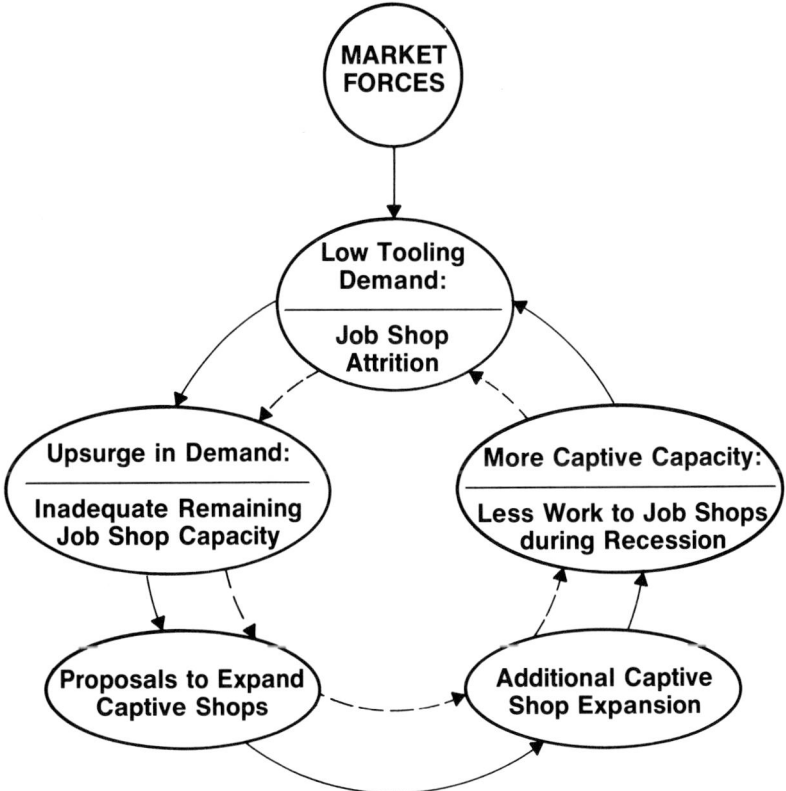

Fig. 9. Relation of cyclical tool-and-die demand to captive shop expansion.

54 / The Tool and Die Industry

shops—a condition which the automakers, themselves, suggest would be intolerable. AMC, with no capability to make major tools, would be at a critical competitive disadvantage. Chrysler could find itself in this position if it continues to dispose of ancillary plants, such as the Warren Tool and Die Plant. The competitive balance of the U.S. auto industry would then hang upon the speed at which a company could introduce appealing styling changes.

NOTES

1. Inflation, as measured by the wholesale price index for metalworking, Apr. 1974, was 136.6 (base year 1967 = 100). It accounted for approximately 40 percent of this apparent growth.

2. General Motors Corporation, *Competition and the Motor Vehicle Industry*. A study submitted to the Subcommittee on Antitrust and Monopoly, Committee on the Judiciary of the United States Senate, and included in the record of the subcommittee's *Hearings* on Apr. 10, 1974 (Detroit, Mich.: General Motors Corporation, 1974), p. vii.

3. *Ibid.*, p. 66.

4. *Hearings* before the Subcommittee on Special Small Business Problems of the Select Committee on Small Business, 91st Cong. 1st Sess. (1969), p. 139.

5. John S. McGee, "Economies of Size of Auto Body Manufacturing," *Journal of Law and Economics*, Oct. 1973: 239–73.

6. *Hearings* before the Subcommittee on Special Small Business Problems...(See Appendix B).

7. During the 1960 decade, the average amount of plastic materials per car grew more than fivefold, from about 15 pounds to 100 pounds. Moreover, one manufacturer increased the amount of plastics from 27 pounds in 1960 to 150 pounds per car in 1975.

8. *Hearings* before the Subcommittee..., p. 151.

9. *Ibid.*, p. 156.

VI

ECONOMICS OF INDEPENDENT TOOL AND DIE SHOPS AND THE MAKE-OR-BUY DECISION

Relative Economics

Assuming that the tool and die work can be done by either captive or independent shops, which is the more economical? In the final analysis, an answer can be given only after comparing one given captive shop with one given independent shop, since some shops in both groups are more efficient than others. Nevertheless, certain generalizations can be made. In this chapter we consider the general characteristics of captive and independent shops affecting relative costs, and also the more specific cost characteristics relating to labor, material, and overhead. In Chapter 7 we make use of this information to decide whether, and when, captive shops should be utilized. There we will also consider arguments based on considerations other than measurable costs in deciding the appropriate utilization of captive shops.

It is axiomatic that a business cannot survive over the long run unless it covers *all* costs and generates a rate of return on the resources employed commensurate with the risks assumed. It also is true, we feel, that divisions of a business should be judged by the same standards. It follows that a business should establish a captive shop for tool and die making, *assuming it is to perform a full range of tooling activities,* only when the full costs of operation plus a reasonable rate of return on the resources employed are equal to or less than what would be paid for the products or services on the outside. This is the standard applied in this chapter and the next in deciding the relative economics of independent and captive shops.

General Characteristics of Captive and Independent Shops which Affect Costs

While captive shops are dependent on one customer, job shops (with the usual exception of the large body shops), though depending heavily on the automotive industry, have a number of customers from other industries. Captive shops are tied to the activity cycle of only one type of business, operating at a low percentage of capacity when this business is operating at a low level and vice versa. Job shops, having as customers a variety of businesses with a variety of business cycles, are often able to operate at a higher average percentage of capacity, thereby reducing the average cost per job performed.

Total labor costs (including fringe benefits) generally are lower and more flexible in independent shops. One reason is that a number of the job shops are not unionized. But this is also true in the smaller unionized shops (60 to 75 workers), where there is more likely to be a spirit of cooperation between management and labor. Commonly, mutual concern for the welfare of the employees and employer is present. Morale and productivity are typically higher. In dealing with smaller numbers, managers have a distinct advantage both in being able to select their employees more carefully and in knowing each worker and his capabilities intimately. Manager and worker appear to be more cognizant that their fortunes are interrelated and interdependent, and that the welfare of each depends on the efforts of the other. Managers are frequently the owners and thus realize that their income is dependent on successful operations. Often they are on hand, working side by side with the employees. Individual officers may divide their time among a number of roles, freqently serving as the purchasing agent, the job estimator, the production supervisor, and the sales manager. Workers realize that their jobs depend on the business operating successfully, and are therefore motivated to do their work expertly.

By contrast, an adversary relationship between labor and management is more likely in larger shops (both independent and captive), with each side suspicious of the other's actions and motives.[1] Because of these attitudes and other, less discernible factors, larger operations less frequently achieve productivity equal to that of smaller organizations.[2]

The supervision and utilization of workers is likely to be more effective in most independent shops because of their smaller size. Supervisors are more aware of employee strengths and limitations because they work more closely with them. Independent shops specialize in design and construction, and the work force deals almost exclusively with a particular type of tool. Thus, the workers develop more refined design and construction techniques and expertise, which enhances productivity. Those who cannot or will not do their jobs properly are released.

This kind of supervision and worker utilization is much more difficult to achieve in captive shops. Captive shops usually furnish a full range of services, such as design, construction, repair, installation, etc. When workers are shifted frequently between these activities, they tend to be less efficient than toolmakers who spend the majority of their time in design or in construction. Perhaps as a testimony to the productivity of job shop workers, employment personnel of the captive shops frequently show preference for any available job shop men when they seek additional work force.

The "30 and out" early retirement that is so popular with the UAW workers appears at least temporarily to have had a negative productivity affect in captive shops. Many of the older, highly experienced and talented toolmakers have chosen to retire, even though they were seven to ten years away from the normal end of their productive careers. In some instances these retirements left vacancies and voids to which less skilled workers were assigned. This undoubtedly has resulted in lower output, at least in the short run.

In certain captive shops there also were situations in which the older, more experienced employees took lay-offs first, under the Supplemental Unemployment Benefits program, when the employee force was reduced during the depressed periods of the last few years. They did this because they had built up larger unemployment benefit increments than newer employees and could receive 95 percent of their base pay for an extended period of time. When these benefits ran out, they bumped the younger, inexperienced workers off the jobs. Whatever might be said for such a procedure on humanitarian grounds, it almost certainly reduced productivity during the periods involved.

Just as there are productivity differences between captive and independent shops, differences also exist among independent shops. A general decline in productivity throughout the industry appears to have occurred over the past ten to fifteen years, but this is more apparent in the larger unionized shops. As mentioned, some closings of large independent body die shops have been attributed to the inability of management to sustain productivity in the face of worker demands for higher wages and benefits and for increased nonproductive time, such as breaks, wash-up time, etc. Surprisingly, managements of more than one defunct large company stated that greater worker paychecks would have been feasible if accommodations could have been negotiated restricting nonproductive time. One manager claimed that productivity had eroded to the point where 15,000 hours were required to build tooling in 1974 that was constructed within 12,000 hours in 1958. Another stated that a job which required 1,000 hours ten years ago presently would take over 2,000 hours. To the extent such experiences are common, and considering the increase in wages and fringe benefits which has taken place during the interval, productivity has declined; it is probably a sizable reduction.

A major difference between independent and captive shops, having both direct and indirect cost effects, is the economic setting in which they operate, particularly regarding those job shops supplying tools and dies to automotive companies as compared with the captive shops of those companies. Such independents operate in a monopsony atmosphere, i.e., a condition of many sellers and few buyers. As a consequence, competition is severe. Jobs are acquired primarily on the basis of prices quoted and buyers have the power to exert considerable influence on such prices. As a result independent shops are forced to attempt to hold the line on increases in wages, fringes, and overhead costs. At the minimum, they must attempt to keep such increases more modest than those in captive shops. The price paid for attempting to do this sometimes has been prolonged strikes at unionized job plants. Yet competition has been the pressure restricting wages and fringe benefits. Such competition has come indirectly from captive shops assuming more and more of the automotive tooling market, but more directly it has come from the nonunionized independent companies operating in less industrialized areas where fringe benefits typically average only about 20 percent of direct labor costs.

On the other hand, captive shops, particularly those established by automotive companies, operate in an industry situated in an oligopolistic setting, i.e., one of few sellers and many buyers. As a consequence, up to some undefined maximum, they can pass cost increases on to the consumer of the product. The pressure to hold the line on costs is not so severe. Considerations other than cost efficiencies often take precedence in decision making. For example, fringe benefits are as great as 42–48 percent of direct labor costs in captive shops, while the corresponding figures total 26–38 percent for independents. Knowing such costs can be passed on to the consumer, labor officials press harder for increases in wage and fringe benefits, while management is less likely to strongly oppose such increases. During the late 1960s and early 1970s, for example, many economists were estimating that between a 2 and 4 percent average increase in wages nationwide would have been sufficient to maintain full employment and absorb additions to the work force, without inflation. Settlements frequently were 8–9 percent, and in some cases ran up to 12–15 percent. The purpose here is not to attach blame. Self-interest is human, will always be present, and is a good trait when self-enlightened and not carried to extremes. When carried to extremes, however, the result can be devastating, as we have seen in the United States and elsewhere in industrialized nations during the past two years.

A point was reached for many goods and services, automobiles certainly were one, where costs pushed prices to the point at which consumers were unable or unwilling to buy. Demand fell and, with it, output. Unemployment rose to the highest level since the depression years of the 1930s. We all suffer in such circumstances, the production worker perhaps most of all. Walter Reuther once said: "The privilege of being a free man in a free society is accompanied by the responsibility of making decisions that are compatible with the well-being of society."[3] Free men are found in corporations, in unions, and in government. They bear a similar burden.

In this regard, another point should be mentioned. Provisions have been written into union contracts at captive shops to prevent work from going outside, or at least making it a strikable issue. A UAW newsletter summarized this development as follows: "The right to strike over violations of the outside contracting provisions of the agreement was won for all workers covered by the contract."[4]

Generally, union officials must be notified in writing before work is let out. They have strongly supported and defended this policy of notification and striking:

> "We are not going to tolerate subcontracting of our work so long as our members—the most skilled in the world—are laid off," said Local 600 President Mike Rinaldi.[5]

Even the Supreme Court of the United States has become involved in such controversies.

> The Supreme Court dealt labor a major setback by expanding the circumstances in which unions can be sued under federal antitrust laws. In a five-to-four decision involving a building-trades union, the high court said unions are subject to antitrust attack if they coerce general contractors into agreements that "indiscriminately" keep nonunion companies from competing for subcontracting work. ... Writing for the five-member majority, Justice Lewis Powell said the resulting agreements "indiscriminately excluded nonunion subcontractors from a portion of the market, even if their competitive advantages were not derived from substandard wages and working conditions, but rather from more efficient operating methods." Mr. Powell said that competition based on efficiency is a goal of antitrust laws and isn't a goal of federal labor policy.[6]

At this stage no one knows what the final result of such controversies will be. There is a strong possibility, however, that labor agreements may well be brought into the fold of the antitrust laws on a recurring basis. The Supreme Court appears interested in promoting competition in all aspects of the economy, including labor activities.

Specific Cost Characteristics

In several respects, evaluations of cost data derived from independent and captive shops during this investigation agree generally with those observed in the 1961 Paton and Dixon study.[7] An important difference in labor costs did materialize, however.

Labor costs

In 1961, Paton and Dixon found that the prevailing labor rates in captive shops were approximately 10 percent lower than those in the independents—$3.42 compared to $3.76 per hour. Depending on the particular labor category, differences in favor of the captive shops ranged from 25 cents to 48 cents per hour. In the mid-1960s the differential shifted in the independents' favor and has widened ever since (see Figure 10). Direct labor rates for so-called bench-mark job categories are compared as follows:

Classification	Captive Shop	Independent Shop*	Difference
Diemaker	$7.33	$7.00	$0.33
Toolmaker	7.27	6.94	0.33
Boring mill operator	7.27	6.94	0.33
Keller operator	7.40	7.08	0.32

*Rates in effect at unionized shops in Detroit Tooling Association in September 1974.

When one considers the total hours required to produce a large body die, the difference in wage rates becomes impressive. Evidently, then, one reason the job shops can compete successfully is because of lower labor costs.

Beyond this additional 33-cents-per-hour wage rate differential now paid by captive shops is an even larger difference in fringe benefit cost. This has resulted from the auto manufacturers granting or enriching such benefits as supplemental unemployment compensation, "30 and out" early retirement, enhanced insurance programs, etc. For example, in 1974 one automotive company paid an estimated $16.9 million as *supplemental* allowances to those who retired early, adding labor costs of about 55 cents per hour during 1974, a sizable burden.

Certainly factors other than labor costs have been involved in the increasing price of cars, such as governmental regulations and requirements, increasing cost of oil in particular and most other raw materials in general, and the general attitude of all who wanted higher pay for less work. Labor must recognize its share of that blame. In the long run, it is not in the best interest of the production worker to increase wages beyond what increases in productivity will support. Certainly it is not in their best interests to decrease productivity. When prices rise so high customers won't buy, the result is layoffs, such as we have seen the past few years, and labor often suffers the most, though all are affected.

Supplemental unemployment benefits alone are running at about 12 cents per hour, a sizable amount considering the total number of hours worked in a captive shop per year. Independent shops do not have this benefit. Data in Table 7 summarize the comparative total labor costs per hour, using the diemaker category as a representative classification.

Comparisons of total direct labor costs and associated fringe benefits for *all* classifications in this study suggest that the costs of independent shops range from approximately $1.00 to $1.39 less than those in captive

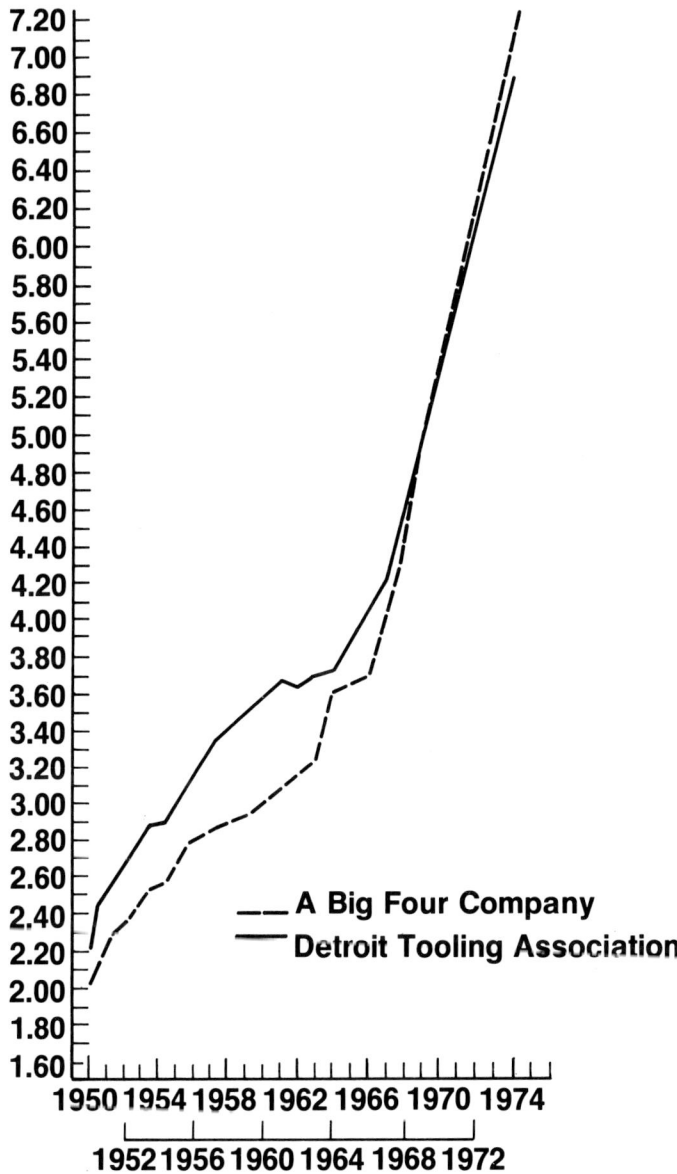

Fig. 10. Diemaker wage rates at captive and independent shops.

Table 7
DIRECT LABOR AND FRINGE BENEFIT COSTS
INDEPENDENT AND CAPTIVE SHOPS

Cost Element	Independent Shop	Captive Shop	Difference
Diemaker hourly rate	$7.00	$ 7.33	$0.33
Fringe benefits as a percentage of hourly cost	26–38	42–48	
Midpoint (percentage)	32	45	
Fringe benefit costs per hour	$2.24	$ 3.30	$1.06
TOTALS	$9.24	$10.63	$1.39

shops. Moreover, this comparison excludes the impact of relative productivity in the two types of shops.

The *range* of hourly wages rates for a given classification in the job shops surveyed is usually very broad as compared with large captive shops. Within these broad ranges the independent shops, in contrast with captive shops, are able to offer an important incentive to labor in the form of merit ratings.

Overhead costs

The total overhead rates of the job shops appear to be surprisingly low. The shop which is devoted solely to the production of tooling has no place to hide its overhead costs; thus the resultant rates are not open to the perennial controversies centering on the question of where to charge a given cost. Total overhead rates as low as 70 percent of direct labor cost are reported, and it is reasonable to assume that rates well under 130 percent are common. In fact, 65 percent of the respondents to the questionnaire reported overhead rates of less than 130 percent, 45 percent reported less than 115 percent. It should be noted that these rates include total manufacturing overhead as well as administrative and selling expenses. Moreover, these rates are comparable to those reported by Paton and Dixon.

> Characteristic all-inclusive rates in job shops range from 70 percent to 130 percent of direct labor cost; comparable rates for large captive shops appear to range from 250 percent to over 400 percent.[8]

The following overhead distribution is typical of the individual components making up the total overhead figure. Overtime premium in this and the following examples is treated as part of the direct labor base rather than as an overhead cost. If the premium were treated as overhead the rates would be increased roughly 20 percent.

Component	Percentage of Costs	
1. Manufacturing		
Supervision		12.12
Indirect labor—wages		
Crane operators	1.40	
Machine repair	4.50	
Sweepers	1.21	
Tool crib	1.72	
Tool maintenance	1.68	
Clerical and other	1.63	12.14
Fringe benefits		
Vacation allowances	3.75	
Holiday payments	2.12	
Pension fund	5.12	
Supplemental unemployment benefit fund	1.76	
Payroll taxes	8.75	
Compensation insurance	1.35	
Group hospitalization	6.95	
Group life and health	2.34	32.14
Other expenses		
Depreciation	8.32	
Perishable tools	2.41	
Equipment maintenance	2.96	
Power, light, heat	4.68	
Tax, corporate state income	1.12	
Tax, other	4.58	
Insurance, fire	.91	
Plant protection	.52	
Trucking and miscellaneous	.26	25.76
Total manufacturing		82.16
2. Estimating		
Salaries	5.23	
Miscellaneous	.75	
Total estimating		5.98
3. Selling		
Salesmen's salaries	2.91	
Other salaries & wages	1.12	
Transportation	.79	
Advertising & customer relations	1.23	
Miscellaneous	.78	
Total selling		6.83
4. Administrative		
Salaries & wages	4.57	
Office supplies & expenses	1.12	
Apprentice training school	.24	

Taxes	.41
Insurance	.32
Total administrative	6.66
TOTAL OVERHEAD	101.63

A sampling of job overhead rates today indicates amounts ranging from 70 percent to 160 percent of direct labor costs, an average of perhaps 115–120 percent. While corresponding overhead rates of large captive shops were not directly ascertained in the course of this study, such evidence as was made available indicates *total* overhead rates ranging from 250 percent to more than 600 percent in a few cases. Thus, overhead rates for both independents and captive shops have increased somewhat since the Paton and Dixon study, but the differential remains about the same. If anything, the advantage has moved a bit more toward the independents.

If direct labor costs of job shops are assumed to average $7 per hour, the combined hourly cost for direct labor and overhead in the reasonably efficient job shops should approximate $14–$16. Quite obviously these rates, as averages, must vary with volume as well as with efficiency. If direct labor costs in large captive shops are assumed to average $7.30 per hour, the hourly cost when combined with the asuumed overhead range of 250 percent to 600 percent may be as low as $25 or as high as $52. It should be emphasized that the range of overhead rates for the captive shops has not been verified by review of particular examples; also, it should be noted that these are intended to be all-inclusive rates (comparable with the job shop examples cited earlier), rather than differential cost, out-of-pocket cost rates, or standard cost rates.[9]

An important factor affecting overhead cost differentials between captive and independent shops is depreciation charges for plant and equipment investments. Typical overhead rates for various types of automotive manufacturing plants generally illustrate the role played by depreciation charges:

Type of Plant	Degree of Extensiveness and Complexity of Plant and Equipment	Percentage* Ranges of Overhead Rates
Press plant	High	600–900
Body assembly	Moderate	300–600
Engine plant	High	700–900
Car assembly	Moderate	300–400

*Of direct labor.

In general, if the operation requires a high capital investment in plant and equipment, the resulting overhead expenses are also high because of the depreciation charges. Unless the direct labor base is comparatively large, the overhead expenses comprise a fairly high percentage of labor costs. An additional characteristic of a modern automotive plant is that, as the capital investment in equipment increases, the direct labor force tends to decrease, thus forcing the relative overhead rate even higher. Despite the direct-indirect cost relationship phenomenon, experiences clearly demonstrate that greater net efficiencies almost always result from increased investments in automotive manufacturing mechanization—if the production volume is large enough. Plant and equipment depreciation expenses affect overhead rates for captive and independent toolmaking operations similarly, although independent shops seem to have controlled this item more effectively. Our observations paralleled those reflected in Paton and Dixon's summarization of the relationship of depreciation to overhead expenses and the importance of these costs in determining the competitive merits of captive and independent shops.

> Depreciation can be an element of considerable proportions in the overhead component of tool and die production costs. The machinery employed is high-priced and, unless used at a fairly high fraction of capacity, can create a loss result in any properly prepared earnings calculation.
>
> To attack this problem the job shops (1) make use of rebuilt equipment wherever practicable, (2) limit their equipment to the capacities and purposes relevant to their fields of specialization, and (3) operate in structures that are consistent with their space requirements.
>
> The impact of more-than-adequate structures on overhead costs has already been mentioned. In brief, it may be repeated that low-cost buildings with ceiling heights at workable minimum levels contribute to overhead cost economy. With respect to some of the large captive shops, it has been said that the excess ceiling heights may well exceed the total cost of a structure built efficiently to house a moderate-sized job shop. It follows, of course, that unused space creates its share of overhead cost; apart from depreciation, in the form of heating, lighting, and maintenance.[10]

Respondents to the questionnaire indicate that independents are still investing large amounts of total capital expenditures in used machinery. They are adept at repairing such machinery and at changing and converting it to the use at hand, frequently at a total cost of 50–60 percent, or less, of new machinery. This reduces total costs significantly, and we could find little evidence that the use of such machinery has reduced the capability of job shops to perform any of the critical tooling operations.

Material costs

Material costs, particularly in the form of tool steels and cast iron, have been increasing dramatically for both independent and captive shops. Prior to 1970, the material cost component of tooling work comprised approximately 10–20 percent of a company's total cost. Now this factor commonly represents 20–25 percent. This dramatic increase appears to have hit the independent and captive shops equally hard.

Nevertheless, captive shops, with their sophisticated and large procurement staffs, are often able to purchase their materials at prices 2–3 percent below those the independents pay. This relationship is not uniform throughout the industry, however. Managers of independent companies, exercising sharp purchasing practices, in many instances obtain material price breaks comparable to those achieved by the captive shops. The less concerned independent manager will likely pay the 2–3 percent premium.

NOTES

1. One case came up in the field survey in which a large body die shop went out of business because of irreconcilable differences with the union. In a number of other cases, union shops were closed and later reorganized and opened as nonunionized shops.

2. Perhaps a word is in order as to the meaning of "productivity." Admittedly, no one as yet has devised a satisfactory method for universally measuring productivity changes. In fact, numerous studies are presently under way by governmental units and other institutions attempting to devise better measurement approaches and techniques. At present, output per man-hour is frequently asserted as an indication of productivity changes. Brief consideration will indicate that this is really a measure of changes in total output, with little, if any, indication of changes in relative efficiency and effectiveness of either the capital equipment or labor utilized. For example, if a completely automated machine were substituted for a manually operated machine and the wage earner were thereby displaced, the present method of calculation utilized would reflect an infinitely large increase in the productivity of that laborer, hardly a realistic conclusion, to say the least. In this study the word "productivity" means changes or comparisons, given a specified time interval and/or a specified type of capital equipment, both assumed held constant for the consideration at hand.

3. Jean de Givry, "Reflections on Major Trends in Labor Relations," *Labor Education* [Geneva, Switzerland: International Labor Organization], Mar. 1968: 11.

4. *UAW Chrysler Newsgram*, Sept. 1973.

5. *Detroit Free Press*, Mar. 27, 1975.

6. "High Court Deals Blow to Labor in Antitrust Issue," *Wall Street Journal*, June 3, 1975: 7.

7. William A. Paton and Robert L. Dixon, *Make-or-Buy Decisions in Tooling for Mass Production*, Michigan Business Reports No. 35 (Ann Arbor, Mich.: Bureau of Business Research, School of Business Administration, University of Michigan, 1961).

8. Paton and Dixon, *Make-or-Buy Decisions . . .*, p. 10.

9. Later we use a 200 percent standard rate to establish a minimum average amount of overhead for captive shops, recognizing that such a figure is unrealistic in terms of full costs.

10. Paton and Dixon, *Make-or-Buy Decision . . .*, pp. 11, 12.

VII

SHOULD CAPTIVE SHOPS BE UTILIZED?

Chapter 6 presented both quantitative and nonquantitative data to compare the economics of captive and independent shops. In this chapter we use these data to suggest whether captive shops ought to be utilized, and if so, when.

Net Cost Impact

Owing mainly to the shift in the direct labor cost differential from the captives' advantage to the independents' and to the enormous costs accruing from the fringe benefits of captive shops, we conclude that the cost relationship between the two classes of plant has shifted even further in the favor of the independents. Paton and Dixon used the following hypothetical example to illustrate their cost comparisons:

Captive Shop
Direct labor, 30,000 hours at $3.45	$103,500	
Overhead (assume 250 percent)	258,750	
Total cost		$362,250
Direct labor, 30,000 hours at $3.45	$103,500	
Overhead (assume 400 percent)	414,000	
Total cost		$517,500

Job Shop
Direct labor, 30,000 hours at $3.75	$112,500	
Overhead (assume 100 percent)	112,500	
Total cost		$225,000[1]

The conclusion Paton and Dixon drew from this example was that "a total cost saving ranging from $137,250 to as much as $292,500 is possible on a 30,000-hour set of dies produced under job shop conditions."

Under present conditions, a third model for captive shops should be considered for those more heavily equipped shops operating at an overhead rate of about 500 percent (full costing basis). Additionally, there are some captive shops that operate on a 200 percent standard overhead rate. (This standard rate may be considerably less than full overhead costs, and probably is, particularly if it is being used to judge efficiency. Nevertheless, we use it here to establish the low side in the comparison.) When these modifications are implemented in the Paton and Dixon example, the following cost relationships result:

Captive Shop
   a. Direct labor, 30,000 hours
      at $7.30 $219,000
      Overhead (assume 200 percent)   438,000
      Total cost $657,000

   b. Direct labor, 30,000 hours
      at $7.30 $219,000
      Overhead (assume 400 percent)   876,000
      Total cost $1,095,000

   c. Direct labor, 30,000 hours
      at $7.30 $219,000
      Overhead (assume 500 percent)   985,500
      Total cost $1,204,500

Independent Shop
   Direct Labor, 30,000 hours at $7.00 $210,000
   Overhead (assume 120 percent)   252,000
   Total cost $462,000

Under these assumed conditions and based on a tooling job requiring 30,000 hours at current expense rates, the independent company's estimated cost advantage would range from $195,000 to $742,500. In contrast to the $15.40 hourly cost of the independent company, comparable rates for the three captive shops in the example are $21.90, $36.50, and $40.15. It is clear from this illustration that hourly costs of captive plants having overhead rates in excess of 300 percent are at least double the hourly cost of the typical independent. Moreover, when productivity differences are considered, an estimated two-to-one cost advantage in favor of the independent companies may be a conservative estimate.

While the projected "high" savings of nearly $743,000 shown in the example would seem to be so large as to raise questions concerning the illustration's validity, we emphasize that examples of cost advantages ranging upwards of 3-to-1 were reported during our survey. Insufficient information was available to scientifically test for the precise amount of the apparent cost advantages purported to be enjoyed by independent companies, but evidence available indicates that the differentials reflected in the example are sometimes on the low side. In one verified case, for example, a set of tools was finished in a captive shop at a cost somewhat in excess of $1 million, after the independent company (which was struck by the union) had purchased the castings and completed 10–15 percent of the construction. The independent had contracted to build the tooling for approximately $500,000. On this particular project, the captive's price for completing the work was almost two and one-half times the independent's contract price, less the materials transferred to the captive plant.

Independent shop managers frequently mentioned similar cost differentials they have encountered. As another example, a left-hand set of body dies was built by an independent shop while the otherwise identical right-hand set was built in a captive shop, requiring twice the hours used by the independent. Such anecdotal examples are difficult if not impossible to evaluate quantitatively, of course, since actual plant cost records of captive shops cannot be inspected. Nevertheless, there is little doubt that a market productivity advantage commonly is enjoyed by smaller independent companies.

From a measurable cost standpoint, therefore, the advantage appears to lie with the independent. There are other factors, however, which ought to be considered in arriving at conclusions as to what are the proper economic functions of captive shops. Certainly these affect costs, and undoubtedly some partially explain the auto manufacturers' acceptance of the cost differentials we have been considering to this point. In most cases, however, it would be difficult, if not impossible, to measure the specific cost impact of any one of them individually or all of them collectively.

Other Economic Considerations

One frequently encountered reason for establishing captive shops is that the company has an unused portion of its plant large enough to house its required tool and die operations. The argument is that this space is idle, the fixed costs related to it are sunk and will continue whether it is used or not. The only costs associated with establishing and operating the tooling activities, then, are the incremental costs associated with the production of those tools and dies, it is alleged. At first blush, this appears to be sound reasoning. In Chapter 4 on pricing policies we emphasized that fixed or sunk costs were irrelevant for making certain types of pricing decisions. Aren't those same conditions present here? The answer is "no," for a number of reasons.

In that chapter we emphasized that pricing below full costs would be appropriate only for short-run decisions which improve the immediate profit picture *without any adverse long-run effects on the business.* The decision to establish a tool and die operation is long-run by its very nature, so the decision must be based on the optimum long-range use of the facilities.

Indirect expenses of captive shops are higher initially because of costs arising from the hierarchial type of organization, accompanied by its administrative controls, reports, and levels of supervision. Unlike most types of modern mass production, in the one-of-a-kind output characteristic of toolmaking operations, these increased hierarchial costs appear to be seldom offset by the greater efficiency gained through such expansion. Despite the general principle that an increased scale of operations (up to a relatively high limit) generally enhances efficiency and reduces costs, toolmaking operations often tend to become less efficient when the organization begins to exceed 75–100 workers. Thus, significantly enlarged toolmaking operations rarely compensate through greater efficiency for the relatively high administrative expenses or central office costs that might be incurred.

A number of factors need to be considered in deciding upon the best use of the space over the long run. If demand for the product increases, the space might be used more effectively in manufacturing operations. Flexibility in planning manufacturing operations and implementing those plans is reduced when the space is devoted to other functions. This is likely to result in increasing costs, since facilities will have to be expanded or new plants acquired when increased production is required. Moreover, the space was undoubtedly designed originally for some purpose other than tooling work. Converting it to this new use might result in less than optimum utilization of the area, with resulting higher overhead costs.

Another cost that should not be overlooked is the substantial outlay of funds necessary for hiring and training employees. It takes years of experience to become a good tool and die maker, and even longer to become a successful manager of a tool and die operation. This experience already has been acquired by those who manage independent firms. It has been alleged that the apprentice programs in union shops are considerably less cost-effective than are those in job shops. The union apprentice program gives a worker 400–600 hours training on each of several different machine tools. In three to four years, after the worker completes his tour of apprenticeship on these various machines, he is assigned seniority status on a machine in a job classification and, for all practical purposes, is restricted to running that machine. Apparently there are certain union rules that discourage the temporary shift to other classifications. Union rules also apparently make it difficult for management to shift workers from one machine to another as the work load changes.

This inflexibility is obviously costly to the plant and presumably frustrating to workers who have been trained to do diverse operations. In addition, an apprentice training on a new machine is far less efficient and proficient in his early stages of training. This increases the cost of production during the training period and adversely affects tolerances and part quality. It also increases the amount of rework necessary to raise the quality to satisfactory standards.

Another important consideration for those thinking about establishing captive shops is the effect their decisions could ultimately have on independent shops. Setting up captive shops naturally reduces the jobs available to independent shops and, if carried far enough, might drive, and in fact has driven, a number of independent shops out of business. Later, if the manufacturing company decides to drop its captive shop operations, it might be difficult to find independent shops which could do the tool and die work. This is not an unrealistic possibility. A number of manufacturing companies have set up captive shops only to discover that it was less costly to have the work done by independent shops. They then had to discontinue this portion of their operations, a change which undoubtedly resulted in losses to the companies involved. Because of the large capital investment required and the skills that are necessary, such shops cannot be brought back into existence easily.

It has been estimated, for example, that the cost today to establish a body die shop equivalent to one with 200–350 employees and adequate buildings and equipment, would be somewhere in the $10–$12 million range. Very few individuals, or small groups of individuals, have that much capital to invest, or even would be inclined to invest that much if they had it, considering the risks involved in such an undertaking. Most of the large independent shops expanded gradually when equipment cost 30–50 percent less than it presently does. There are also hidden costs to keep in mind. One of these is the interest which could be earned on the money in another investment. Assuming a 10 percent rate of return, $12 million would earn $1.2 million per year. Property taxes would amount to approximately $350,000 per year, to say nothing of payments for insurance, security, and maintenance. Multiply these by five, six, or perhaps ten to reflect the total tooling capacity that might be required to satisfy the needs of the Big Four should all large independent body die shops go out of business, and additional total costs of $8–$10 million per year for the customers are easy to visualize.

The sunk-cost syndrome mentioned earlier apparently also has encouraged customers of the independent tool and die shops to take the position that they will do a substantial amount of their own tool and die work unless they are able to have this work done outside at prices generally comparable to estimated differential or out-of-pocket costs in an existing, well-equipped captive shop. Taking advantage of bargain or "distress" prices in particular situations for short periods is no crime, but it is not reasonable to treat such special circumstances as normal bench marks in deciding whether to buy or make. Independent job shops, like any other enterprise, cannot stay in business for long by following a policy of bidding for work at less than its full costs, including all overhead charges. Full cost to the job shop generally will be higher than the estimated incremental cost to the captive shop, despite the cost advantages inherent in the efficient job shop. Moreover, it must be remembered that even full cost is not a proper measure, for any considerable period, of selling price. In a competitive market economy, the well-run job shops must find it possible to make a return on the capital employed commensurate with the risks assumed or resort to a process of liquidation, either abruptly or gradually. We are sure most customers recognize this from the vantage point of their own operations, so it is difficult to understand how they can expect other producers to be satisfied for long with a mere break-even operating result.

There is ample evidence that private funds will not flow into, or remain in, any field of business which does not provide at least a good prospect of yielding an attractive return on the resources employed, in light of the risks assumed. The question from the viewpoint of captive shops, therefore, is whether they can produce tooling at a total cost (full costs plus a capital-attracting rate of return) equal to or less than it can be acquired from the outside. As stated earlier, when captive shops are created, that decision, by its very nature, is a long-run one. Plant capacity, as well as the machinery and equipment, will have to be replaced at some future date. Unless this "full cost plus a return on investment" criterion is used, the owners of the manufacturing plants will not gain full advantage of long-run profit maximization, and the customer will bear the burden of higher prices. If captive shops operate on the theory of out-of-pocket costs, they should realize that their production facility is being liquidated. This is true since the price charged is insufficient to cover depreciation charges, and the consequent funds generated by operations are insufficient to replace the capital equipment when it wears out.

The investment in unfinished work

Outside purchasing of tooling has other financial advantages. The use of job shops can reduce the customer's capital investment needs by many millions of dollars. Except for tooling purchased from foreign sources, it is not common in this industry for the customer to make payments in advance of final delivery. The suppliers thus make possible the release of many millions of dollars of working capital on a year-round basis. A given customer, for example, may require tooling costing well over $100 million. Avoiding investment in unfinished work for the duration of such a program constitutes a clear-cut and significant financing by the supplier. In this regard, however, it should be pointed out that some alterations need to be made. After so many "loss years," independents are finding it increasingly difficult to obtain and maintain sufficient working capital to finance major programs.

Tools have become bigger and more complex, and they take longer to build. With this, prices of most all cost components, including interest, have soared. The large customers (such as automobile producers) frequently can borrow funds at a more favorable rate than the job shops. Since the cost of borrowing, as well as other costs and the rate of return, must be covered by the selling price, the total price to the customer would be lower over the long run if progress payments were made in some fashion to the independents. Many of the foreign tooling purchases made by the Big Four have provided for progress payments, and progress payments are now common in other domestic operations, such as the construction industry. There risk is assumed by both buyer and seller—a reasonable approach considering the large amounts of capital involved—although provisions are usually made to hold some reasonable amount in escrow as protection for the buyer. One possibility is described below, although others might be considered. The customer might pay for designs immediately upon approval and material upon delivery to the plant. For example, 25 percent of the remaining contract price could be paid when the work is 50 percent completed, an additional 25 percent when it is 75 percent done, and the remaining 25 percent after all work is completed and accepted, according to the contract terms.

In concluding this section, we emphasize that previous data indicate that captive shops probably cannot successfully compete with independent shops in the production of dies on a full cost-plus-rate-of-return basis. If competitive job shops are essential to the larger users of tools and dies, such as automobile producers (this seems to be the prevailing opinion of the managements of these producers as indicated by comments at the congressional hearings reproduced in Appendix B, and in discussions during our field survey), current findings indicate the utmost importance of large manufacturers' carefully reconsidering the make-or-buy decision.

We should also point out, however, that we have been considering the question of whether captive shops should be utilized as a full and complete substitute for independent shops. The question remains: Are there any functions captive shops can perform more economically than independent shops? The answer is yes, and the following section describes those circumstances.

Economical Functions of Captive Shops

Skilled craftsmen and efficient machinery near the production lines are a necessity to maintain the dies as needed. For this reason, stamping plants establish tool and die rooms; this probably represents a minimum requirement for them. Such work could be performed by independent shops, but the production losses that would result by moving the dies in and out of the plant plus the direct costs involved would make the total expenses prohibitive. Equally important, the manufacturer of stampings (and the user of dies) requires a crew of skilled hands for the adaptation and installation of new dies as they are received from outsiders and are placed in the production presses. In some instances the installation period discloses ways of modifying the press operations for the sake of economy or of improved product, and the dies are given further processing or modification. Finally, dies sometimes need to be changed to meet production changes and alterations. For all of these reasons, minimum captive tool and die shop facilities are a necessity. Relying on the outside supplier for these services, while a possibility, would undoubtedly not prove economical to the manufacturer.

Some *reasonable* amount of capacity to insure supply is also justified. With the loss of a number of major independent body die shops, there probably is insufficient capacity on the outside to take care of the complete needs of all the customers, particularly considering that much of the work would be let at one time. In other words, this is insurance against two types of risk: first, the risk that a number of large companies may all decide in the same year to make major or complete model changes in a substantial part, if not all, of their product lines. The tooling demand that would result from such a coincidence could swamp the job shop facilities, prevent fulfillment of manufacturing commitments and, in a sense, put the manufacturers at the mercy of the relatively small, independent shops. This would be particularly true if the independent shops (or a significant number of them) were unionized and a strike took place at this time. Captive tooling shops thus are seen as a form of insurance against coincident industry peaks and, perhaps, strikes. Second, there exists the risk that, even after a substantial part of the uncomfortably short lead time has passed, it will become apparent that the projected product, as presently designed, might flop in the face of perceivable competition, and a crash program of change would be required. Again some believe that internal facilities under the complete control of management are needed as an insurance factor.

We should mention here that a die cannot be completed in a relatively shorter amount of time simply by adding men to the job, at least not without total costs increasing substantially. Presently, the insurance facilities only provide the owner with the unrestricted power to start tooling work that might otherwise have to wait its turn (this assumes that captive facilities either have unfilled capacity or are occupied with less essential work that may be shelved temporarily).

One or two points need to be made in this regard. Collectively independents could do a better job of satisfying customer needs on a timely basis and thereby reduce this insurance requirement, if sufficient lead time were allowed by the customers. Otherwise, job shops would have to deliberately plan for, and have available, sufficient idle capacity to fill any such requests or go into overtime situations, with the attendant higher costs.

Job shop managers also must have sufficient time to plan and organize their production activities. To accomplish this and to prevent the problems created by unreasonable time demands, job shop owners should get together with their major customers at least and discuss tooling requirements as far into the future as is feasible. This would allow them to plan and set up production schedules in a more orderly fashion, at least for the major jobs. Not only would this procedure foster more favorable relations with customers, but it also would prevent some unnecessary losses for both the independent shop and the customer. It would assure the customer of getting the work when needed and prevent his operations from being slowed down or stopped. Proper planning in this area is likely to be less expensive to the customer in the long run than if he expanded his own facilities to include tool and die making, particularly when these facilities might be idle a substantial part of the time, solely to ensure adequacy of supply on a timely basis.

It should also be noted that if this insurance criterion is carried to the ultimate degree all work would be done internally and all independents would be driven from the scene, with the consequences we have already discussed. Certainly there must be an optimum point or range reflecting the balance between captive and independent facilities, and the rate of demise of job shops in the past ten to twelve years suggests that the trend toward captive shops has outrun its logical course in providing a reasonable reserve.

A further reason for internal production of tooling may come from the desirability of full utilization of expensive, technologically advanced equipment, such as numerical control. While the job shops as a whole possess a full complement of modern equipment, their individual capital resources do not allow speculative investment in high-cost equipment which is unproved and cannot currently be used for normal construction of special tooling.

In conclusion, captive shop capacity is needed to provide maintenance and repair services, to insure adequate availability of research facilities, and to provide a reasonable reserve for assurance of supply. Exactly what their total capacity should be to accomplish these objectives is difficult to say. Evidence does indicate, however, that captive shops have expanded well beyond that point.

NOTE

1. William A. Paton and Robert L. Dixon, *Make-or-Buy Decisions in Tooling for Mass Production,* Michigan Business Reports No. 35 (Ann Arbor: Bureau of Business Research, School of Business Administration, The University of Michigan, 1961), p. 17.

VIII

CONCLUSIONS AND RECOMMENDATIONS

The long-term outlook for the tool and die industry, in the aggregate, is fairly good. The growth rate for the next five years is expected to be somewhat similar to growth during the 1965–70 period. Near-term demand for automotive sheet metal tooling, on the other hand, is not good. The market is expected to be as variable as, and at the relatively low depressed levels of the 1970–74 period. There probably will be one peak year, perhaps the 1978 model year.

We found few who dispute that independents produce dies for less than captive shops when the latter use the full costing approach. The major cost differentials arise in fringe benefits, which cost independents about 50 percent less than captives, and in overhead costs which are 70–160 percent of direct labor costs for independents as compared to 250–600 percent for captives. However, we do not expect the cost advantage to have much of an impact on make-or-buy decisions for automotive sheet metal dies during most of the years during 1975–1979. This is a result of the relatively low demand General Motors, Ford, and Chrysler anticipate for these tools. The demand is not expected to exceed potential in-house capacity by much. Fundamentally, this condition results from the substantial expansion of die construction capacity in the Big Three during the 1960s, and also because of the substantial decline in automotive styling changes. Furthermore, there is a constant move to standardize major body and quarter panel components, thereby greatly increasing their interchangeability between different models.

While the cost savings inherent in producing tooling in independent shops probably will not affect make-or-buy decisions in most of the next five years, awareness of these savings is expected to restrict any further substantial expansion in captive shops. There likely will be some shrinking—particularly at Chrysler where some complete installations are expected to be liquidated, and through attrition at Ford and General Motors, where the tool construction staff is not expected to be rebuilt to their 1968–1970 peaks. Presently it looks as though a peak tooling demand will not be sustained for longer than a model program; thus the auto companies may increase staff for these temporary peaks and assume the Supplemental Unemployment Benefit obligations only when the independent companies have died out to the extent that they cannot handle the overload.

The demand for large sheet metal dies during 1975–80 probably will not fall below the disaster levels of 1974–75, however. It is possible, therefore, that those large sheet metal die companies which were able to weather this storm, i.e., either break even or incur only small losses, will survive under the market conditions that will characterize the next five years. In most of these years they probably will not have large profits, but they may be able to survive and make sizable profits if peak years occur.

Companies that restrict their business to large sheet metal dies and are able to survive will need to
  1. Minimize fixed costs.
  2. Eliminate so far as possible the variability of their business potential by serving new markets. This will necessitate
     a) Improving worker productivity
     b) Paying costs and fringe benefits comparable to those paid by tool and die companies in other parts of the country rather than those paid by the Big Four
     c) Avoiding costly specialized equipment useful to only one industry customer
     d) Increasing their marketing know-how and penetration to nonautomotive, non–sheet metal applications.

The large sheet metal die shops that continue in business also should consider the business opportunities available in plastic molds. They should acquire a design capability arrangement that will enable them to build the molds, either by doing their own design or by subcontracting it out to reliable sources.

Tools imported from Europe are no longer cheaper than tools from domestic sources. If the Europeans begin to increase their U.S. market share, it may indicate dumping. Should these imports grow dramatically, it could signal a potentially serious problem for domestic firms.

The definite and potentially serious shortage of skilled tool and die makers is masked by the present economic depression. This shortage is likely to be critical across the tool and die industry when the economy recovers.

The nature of the firms within the tool and die industry and the changing character of the problems they face are probably the major reasons why it is difficult to solve their problems satisfactorily. Many of the firms are small, and the costs of budgeting, research, and accounting analyses that are adequate for efficient and effective operations, may be prohibitive for them individually. The main characteristic of the problems facing the industry is that in many cases the problems are not peculiar to the individual firm; i.e., they are regional or national in scope.

A few years ago this was not so true. Problems were generally confined to a particular firm, and they could be solved by the management of that firm. At present, when the most pressing problems are industry-wide or regional in scope, the search for solutions must be approached on a broad front. A great deal of cooperation among the individual firms will be necessary to accomplish this.

One possibility, which has been used successfully in other instances, is for the firms of a given area to join in forming a foundation. (If certain requirements are fulfilled, it could be a nontaxable institution.) The foundation would be financed by investments of the individual firms apportioned on some equitable basis determined by the owners. The major purposes of the foundation would be to:

1. Centralize research activities toward developing new and better tooling methods. Technological advance is essential to the future success of tool and die firms.
2. Perform centralized accounting services, such as providing pricing and cost analyses, answering questions on financial problems, and developing and disseminating accounting procedures and techniques.
3. Conduct market research and promotional activities. (For example, the entire tool and die industry can be brought to the attention of the public through general advertisement.)
4. Act as a training center for foremen, managers, and technicians, where programs could be directed primarily at problems peculiar to the area. For example, budgeting techniques and pricing policies could be discussed.

There are a number of advantages to this arrangement. All participating firms could have the advantages of a research facility. This access would be especially helpful to those firms which are unable to afford any individually conducted research or sufficiently extensive research to be effective in their operations. By pooling their resources, firms could also make more attractive offers to college graduates in the scientific and financial areas than are possible at present. Furthermore, the services of these people could be used effectively and efficiently, whereas in many cases there would not be enough work in individual firms to warrant hiring them.

The idea of mergers between some firms should be looked into very carefully. We have discussed the disadvantages of cutthroat price competition both to the tool and die maker and the customer. Independent firms should start competing more on the basis of quality and service and less on the basis of price, and the Big Four, within reasonable limits, should select tooling sources on the same basis. Within the industry *as a whole* changes in prices are not likely to affect volume materially, since volume is largely fixed by the needs of the customers. Even if the price of tooling has a substantial effect on the price of the finished product, cutthroat competition results only in short-run benefits to the customer and irreparable damage to specific tool and die firms.

However, even if such competition is completely eliminated, certain areas of the country still may have substantially more tool and die companies of a given specialization than are justified by the volume of work available, judging on the basis of profits. Here consideration definitely should be given to mergers.

Possible advantages of mergers

There are a number of advantages possible from mergers. (1) If one company has had a period of loss years and merges with a company which has been profitable according to accounting methods prescribed by the Internal Revenue Service, the losses can be carried backward, forward, or both, to offset the profits. In many cases this results in substantial tax refunds which can be used to improve working capital, to expand, and to modernize.

2. The increased size of the firm after merger may make it possible to buy materials in larger quantities. This may result in substantial savings, since the purchase price of materials frequently depends on the quantities purchased.

3. Managerial abilities are combined in a merger. Management tasks can be allocated according to the managerial strengths of the persons involved.

4. Competition will be lessened by merger, making it possible to charge prices more in line with actual market conditions. As indicated previously, this is an advantage to both the customer and the independent shop in the long run.

5. The new company will probably operate at a higher percentage of capacity, thereby utilizing fixed facilities more efficiently.

6. Old items of machinery and equipment that are no longer needed can be sold. Not only does this mean that the most modern equipment of each of the combining units is retained, making the new firm as a total unit more productive, but the funds from such sales can also be used to improve working capital, to expand, and to modernize.

7. In most firms a certain minimum number of employees must be maintained regardless of the percentage of capacity of operations. Mergers often make it possible for a company to employ fewer workers than its predecessor companies combined. As a result, the less efficient employees need not be kept.

Mergers present complex problems and should be undertaken only after careful consideration and counsel by experts in this area. However, they have enough attractive possibilities that they should be given careful attention. It should be remembered that there are many types of mergers. Merger does not only mean an actual combining of physical operations. Even if it is decided that physical combinations are not feasible or desirable, gentlemen's agreements, referral activities, and many other mutual arrangements can have beneficial effects.

Another possible form of cooperation is to develop pooled arrangements with colleges and universities in establishing and operating educational and research programs.

Cooperation in the industry

Though it is easy to say, and difficult to do, management and labor, independents and captives, must work more closely together, if any are to survive economically over the long run. The history of the past few years may be the beginning and continuation of a trend toward the ultimate disabling of what is an invaluable segment of the tool and die industry, i.e., the large independent body die shops. Almost unanimously, those in the captive shops (and managers of firms owning captive shops) to whom we talked during the survey, stated that the capacity of these shops is indispensable at various times. That capacity cannot survive without cooperation, and some of the Big Four may not be able to survive the future without the independents' capacity.

Admittedly, a part of the decline in independent shops lay with certain job shop managers themselves. Some are inept in price determination practices; some accepted jobs which could not be delivered on time; some simply were not good managers. Most such shops probably are no longer in business, but improvements in overall management techniques could be made by following the recommendations stated earlier in our report. We are optimistic that such improvements will take place, particularly considering the competitive pressures.

Captive shops are indispensable for maintenance of production tooling. They are not an economical replacement of the job shop pool; nor can a proper balance between the two persist where the job shops find it necessary to sell at distress prices continuously. An analysis of representative job shop labor and overhead costs, and the conditions under which shop operations are managed, implies strongly that in the tool and die business the small, privately-owned shop operates at substantially lower costs than does the large captive shop. Tools and dies simply cannot, at the present stage of the arts, be produced more rapidly, more efficiently, or more cheaply simply by putting more men on the job, nor can they ever be made more cheaply simply by excluding an important portion of total cost from consideration.

It is our belief that tool production in large captive shops may well be justifiable economically as fill-in work in maintenance shops, as regular work in connection with research programs, and as regular work where insurance of supply must be provided at all costs. At the same time, the total long-run costs of such operation should not be ignored. It is strongly doubted that tooling production in such shops can meet the

Conclusions and Recommendations / 81

competitive test of full costing. It is therefore concluded that, in the interest of economical operation, the realm of activity of captive shops should be held within the confines of the maintenance, fill-in, research, and vital insurance requirements.

Finally, we would like to emphasize that the existing problems need not portend a dismal picture for the future of the independent tool and die shops. The tooling industry is not peculiar in having problems; all industries have them. They must be attacked with vigor, and long hours and much effort will be required before adequate solutions are found. If this course is followed, and assuming the necessary cooperation of the customers is forthcoming where needed, we see a bright future for the industry.

APPENDIX A

STATISTICAL TABLES

Table A-1

VALUE OF SHIPMENTS AND EMPLOYMENT IN THE TOOL
AND DIE INDUSTRY (SIC 3544) FOR 1947–72

Year	Value of Shipments (Millions of Dollars)			Employment (in Thousands)		
	Original	1954 Adjustment (Less 33.9 Percent)	1958 Adjustment (Plus 17 Percent)	Original	1954 Adjustment (Less 36.7 Percent)	1958 Adjustment (Plus 17.9 Percent)
1947	$ 661.3	$ 437.1	$ 511.5	95.9	60.7	71.6
1949	628.9	415.7	486.4	81.5	51.6	60.8
1950	825.5	545.7	638.4	87.2	55.2	65.0
1951	1,250.4	826.5	967.0	112.4	71.1	83.9
1952	1,514.0	1,000.8	1,170.9	124.5	78.8	92.9
1953	1,708.5	1,129.3	1,321.3	137.4	87.0	102.5
1954	931.5		1,089.9	77.4		91.3
1955	(N.A.)†			82.7		97.5
1956	(N.A.)			91.8		108.2
1957	(N.A.)			92.2		108.6
1958	1,060.6			83.3		
1959	1,235.0			89.4		
1960	1,315.0			93.2		
1961	1,201.9			90.4		
1962	1,488.3			101.2		
1963	1,388.8			90.9		
1964	1,571.3			93.8		
1965	1,839.2			108.5		
1966	2,217.9			120.3		
1967	2,202.3			113.6		
1968	2,193.8			106.7		
1969	2,387.7			114.5		
1970	2,470.8			111.8		
1971	2,160.3			98.8		
1972	1,792.0			97.6		

Source: U.S., Dept. of Commerce, Bureau of the Census, *Census of Manufactures* and *Annual Survey of Manufactures* (Washington, D.C.: Government Printing Office, appropriate years).

Note: The original figures for 1947–53 are for the cutting tools, jigs, fixtures, etc., industry group (old SIC 3543). The 1954 amendment by the Bureau of the Budget to the Standard Industrial Classification divided old SIC 3543 into the special dies and tools industry group (old SIC 3544) and the metal working machinery attachments industry group (old SIC 3545). The original figures for 1954–57 are for old SIC 3544. The present special dies and tools industry group (SIC 3544) is the result of the 1957 Standard Industrial Classification revision which again resulted in a new definition of this industry. These are the original figures which appear for 1958–72.

For adjusted figures, we reviewed the composition of the new industry and its relationship to the old industry at the time an amendment was made. Original figures are not comparable; the figures adjusted in the 1958 column are somewhat comparable. Data are not available for 1948.

*Value of shipments for the old industry in 1958 was not published, so 17 percent was selected.

†N.A. = not available.

Table A-2
VALUE ADDED BY MANUFACTURE FOR THE TOOL AND DIE
INDUSTRY (SIC 3544) AND GROSS NATIONAL PRODUCT 1947–72

Year	Value Added by Manufacture (Millions of Dollars)			Gross National Product (in Billions of Current Dollars)
	Original	1954 Adjustment (Less 33.4 Percent)	1958 Adjustment (Plus 17.1 Percent)	
1947	$ 502.0	$334.3	$ 391.5	$ 231.3
1949	457.4	304.6	356.7	256.5
1950	609.6	406.0	475.4	284.8
1951	889.6	592.5	693.8	328.4
1952	1,103.8	735.2	860.9	345.5
1953	1,218.7	811.6	950.4	364.6
1954	682.1		798.7	364.8
1955	686.4		803.8	398.0
1956	935.2		1,095.1	419.2
1957	856.0		1,002.4	441.1
1958	780.1			447.3
1959	932.0			483.7
1960	966.5			503.7
1961	901.5			520.1
1962	1,106.6			560.3
1963	1,029.3			590.5
1964	1,180.1			632.4
1965	1,341.7			684.9
1966	1,664.5			749.9
1967	1,646.7			793.9
1968	1,631.4			864.2
1969	1,775.9			930.3
1970	1,800.7			977.1
1971	1,556.7			1,055.5
1972	1,792.0			1,155.2

Source: U.S. Department of Commerce, Bureau of the Census, *Census of Manufactures* and *Annual Survey of Manufactures* (Washington, D.C.: Government Printing Office, appropriate years) for value added by manufacture. U.S., Department of Commerce, Office of Business Economics, *Business Statistics: 1973* (Washington, D.C.: Government Printing Office, 1973), p. 1, for GNP figures.
Note: For explanation of method used to derive the 1958 adjusted figures, see Note, Table A-1.

Table A-3

CONVERTING GROSS NATIONAL PRODUCT AND VALUE ADDED BY MANUFACTURE FOR SPECIAL DIES AND TOOLS TO CONSTANT DOLLARS AND PERCENTAGE OF GROWTH: 1947–72 (1967 = $1.00)

Year	Purchasing Power of Consumer Dollar	GNP (in Billions of Constant Dollars)	Percentage of Growth	Value Added (in Millions of Constant Dollars)	Percentage of Growth
1947	$1.495	$345.8	—	$ 585.3	—
1949	1.401	359.4	3.9	499.7	−14.6
1950	1.387	395.0	9.9	659.4	31.9
1951	1.285	422.0	6.8	891.5	35.2
1952	1.258	434.6	3.0	1,083.0	21.5
1953	1.248	455.0	4.7	1,186.1	9.5
1954	1.242	453.1	− .4	992.0	−16.4
1955	1.247	496.3	9.5	1,002.3	1.0
1956	1.229	515.2	3.8	1,345.9	34.3
1957	1.186	523.1	1.5	1,188.8	11.7
1958	1.155	516.6	−1.2	901.0	−24.2
1959	1.145	553.8	7.2	1,067.1	18.4
1960	1.127	567.7	2.5	1,089.2	2.1
1961	1.116	580.4	2.2	1,006.1	− 7.6
1962	1.104	618.6	6.6	1,221.7	21.4
1963	1.091	644.2	4.1	1,123.0	− 8.1
1964	1.076	680.5	5.6	1,269.8	13.1
1965	1.058	724.6	6.5	1,419.5	11.8
1966	1.029	771.6	6.5	1,712.8	20.7
1967	1.000	793.9	2.9	1,646.7	− 3.9
1968	.960	829.6	4.5	1,566.1	− 4.9
1969	.911	847.5	2.2	1,617.8	3.3
1970	.860	840.3	− .8	1,548.6	− 4.3
1971	.824	869.7	3.5	1,282.7	−17.2
1972	.799	923.0	6.1	1,431.8	11.6

Note: Source for consumer dollar value was U.S., Department of Commerce, Bureau of the Census, *Statistical Abstract of the United States: 1974* (Washington, D.C.: U.S. Government Printing Office, 1974), p. 404. This value was multiplied by current dollars presented in Appendix A, Table A-2, to compute constant dollar values.

Table A-4
REGIONAL DISTRIBUTION OF TOOL AND DIE ESTABLISHMENTS: 1972, 1967, 1963, 1958, and 1954

Region	1972		1967		1963		1958		1954	
	Number	Percentage	Number	Percentage	Number	Percentage	Number	Percentage	Number	Percentage
East North Central	3,236	48.9	3,258	49.2	2,834	48.1	2,819	49.1	2,583	49.6
Middle Atlantic	1,264	19.1	1,340	20.3	1,246	21.1	1,269	22.1	1,227	23.5
New England	648	9.8	713	10.8	682	11.6	631	11.0	581	11.2
Pacific	586	8.9	545	8.2	536	9.1	518	9.0	355	6.8
All other regions*	882	13.3	759	11.5	598	10.1	508	8.8	463	8.9
	6,616	100.0	6,615	100.0	5,896	100.0	5,745	100.0	5,209	100.0

Source: U.S., Department of Commerce, Bureau of the Census, *Census of Manufactures: 1972*, MC72(2)–35C; *Census of Manufactures: 1967*, p. 35C–9; *Census of Manufactures: 1963*, pp. 35C–8 to 9; *Census of Manufactures: 1958*, p. 35C–7, and *Census of Manufactures: 1954*, pp. 35B–4 to 5, for adjusted 1954 figures (Washington, D.C.: Government Printing Office, appropriate years). (See first footnote in Appendix A, Table A–1, for description of adjustment.)

*Includes West North Central, South Atlantic, East South Central, West South Central, and Mountain.

Table A-5

LEADING STATES ACCORDING TO VALUE ADDED BY MANUFACTURE
IN THE TOOL AND DIE INDUSTRY: 1972, 1967, 1963, and 1958

State	1972		1967		1963		1958	
	Amount (in Millions of Dollars)	Percentage of U.S. Total	Amount (in Millions of Dollars)	Percentage of U.S. Total	Amount (in Millions of Dollars)	Percentage of U.S. Total	Amount (in Millions of Dollars)	Percentage of U.S. Total
Michigan	$ 535.1	29.7	$ 518.3	31.5	$ 338.6	32.9	$260.6	33.4
Ohio	243.6	13.5	231.2	14.0	144.2	14.0	97.4	12.5
Illinois	168.3	9.4	158.0	9.6	91.9	8.9	69.5	8.9
California	103.8	5.8	104.7	6.4	70.8	6.9	53.4	6.8
New York	89.3	5.0	83.7	5.1	59.4	5.8	50.4	6.5
Indiana	94.7	5.3	89.1	5.4	54.8	5.3	43.7	5.6
Total for selected states	$1,234.8	68.6	$1,185.0	72.0	$ 759.6	73.8	$575.0	73.7
Total U.S.	$1,799.4	100.0	$1,646.7	100.0	$1,029.3	100.0	780.1	100.0

Source: U.S., Department of Commerce, Bureau of the Census, *Census of Manufactures: 1972*, MC72(2)-35C; *Census of Manufactures: 1967*, p. 35C-9; *Census of Manufactures: 1963*, pp. 35C-8 to 9, and *Census of Manufactures: 1958*, p. 35C-7 (Washington, D.C.: Government Printing Office, appropriate years).

Table A-6

DISTRIBUTION OF EMPLOYMENT BY FIRM SIZE IN THE
MICHIGAN TOOL AND DIE INDUSTRY
EXCLUSIVE OF CAPTIVE SHOPS

Employment Range	Number of Firms	Percentage of All Firms	Number of Employees	Percentage of All Employees	Mean Number of Employees per Bracket of Firms
0- 3	166	15.6	457	1.7	3
4- 9	271	25.5	1,726	6.3	6
10- 19	289	27.1	3,961	14.5	14
20- 49	228	21.4	6,734	24.6	30
50- 99	73	6.9	4,885	17.8	67
100-249	28	2.6	4,122	15.1	147
250+	10	0.9	5,483	20.0	548
Total	1,065	100.0	27,368	100.0	26

Source: Michigan Employment Security Commission, Lansing, Michigan, unpublished data, First Quarter 1974.

Statistical Tables / 89

Table A-7

VALUE ADDED PER PRODUCTION WORKER MAN-HOUR FOR
TOOL AND DIE AND SELECTED CUSTOMERS' INDUSTRIES
(CONSTANT DOLLARS, 1967=$1.00)

Year	Tool and Die (SIC 3544)	Computers and Related Machines (SIC 3573,3574)	Household Laundry Equipment (SIC 3633)	Metal-Cutting Machine Tools (SIC 3541)	Motor Vehicles and Parts (SIC 3711, 3712,3714)
1947	$4.68	$ 4.05	$ 4.98	$ 4.52	$ 4.82
1949	4.72	4.49	5.80	5.13	5.84
1950	5.39	4.95	6.15	4.96	6.43
1951	5.19	4.68	6.13	4.86	5.97
1952	5.72	4.97	6.44	6.49	6.40
1953	5.80	5.26	6.27	7.12	6.27
1954	5.73	5.69	7.29	6.93	6.53
1955	5.55	5.50	8.23	6.73	8.22
1956	6.20	6.73	8.73	7.35	8.21
1957	5.77	6.86	9.00	7.54	8.74
1958	6.44	6.79	10.47	6.99	8.82
1959	6.75	8.41	10.61	8.02	9.72
1960	6.67	7.70	9.56	8.93	9.75
1961	6.44	8.49	11.52	8.44	10.16
1962	6.61	9.59	12.50	9.12	11.13
1963	6.63	10.06	13.41	8.82	10.46
1964	7.09	14.73	12.46	9.25	10.51
1965	6.78	16.04	12.95	9.72	10.89
1966	7.39	17.16	11.89	10.26	10.54
1967	7.69	15.02	12.12	10.56	11.53
1968	7.84	15.69	13.07	11.00	12.18
1969	7.60	15.56	13.43	11.32	11.97
1970	7.89	16.42	12.91	10.26	11.14
1971	7.73	16.63	14.78	10.47	13.54
1972	8.23	18.63	14.83	10.56	13.04

Source: U.S., Department of Commerce, Bureau of the Census, *Census of Manufactures* and *Annual Survey of Manufactures* (Washington, D.C.: Government Printing Office, for appropriate years).

Table A-8

BENCH-MARK DATA FOR TABLE A-7

	Tool and Die (SIC 3544)* Man-Hours (in Millions)		
Year	Original	1954 Adjustment (Less 32.2 Percent)	1958 Adjustment (Plus 17.9 Percent)
1947	156.5	106.1	125.1
1949	132.3	89.7	105.8
1950	153.0	103.7	122.3
1951	214.7	145.6	171.7
1952	236.9	160.6	189.4
1953	255.9	173.5	204.6
1954	146.9		173.2
1955	153.1		180.5
1956	184.1		217.1
1957	174.6		205.9
1958	139.9		
1959	158.2		
1960	163.3		
1961	156.3		
1962	184.9		
1963	169.3		
1964	179.1		
1965	209.3		
1966	231.8		
1967	214.2		
1968	199.8		
1969	212.9		
1970	196.3		
1971	165.9		
1972	173.9		

Continued

Source: U.S., Department of Commerce, Bureau of the Census, *Census of Manufactures* and *Annual Survey of Manufactures* (Washington, D.C.: Government Printing Office, for appropriate years).

*Value added by manufacture (in millions of constant dollars) is shown in Table A-3. Man-hours for the old industry in 1958 were not published, so 17.9 percent was selected as the 1958 adjustment.

Table A-8 *(Continued)*
BENCH-MARK DATA FOR TABLE A-7

	Metal-Cutting Machine Tools (SIC 3541)		
	Value Added by Manufacture (in Millions of Dollars)		Man-Hours (in Millions)
Year	Original	Constant Dollars**	Original
1947	$ 343.2	$ 513.1	113.4
1949	272.7	382.1	74.5
1950	316.7	439.3	88.6
1951	588.9	756.7	155.7
1952	1,003.3	1,262.2	194.6
1953	1,000.0	1,248.0	175.3
1954	743.6	923.6	133.2
1955	691.0	861.7	128.1
1956	882.2	1,084.2	147.5
1957	797.5	945.8	125.4
1958	421.0	486.3	69.6
1959	505.4	578.7	72.2
1960	578.2	651.6	73.0
1961	550.1	613.9	72.7
1962	671.0	740.8	81.2
1963	699.3	762.9	86.5
1964	836.2	899.8	97.3
1965	993.6	1,051.2	108.1
1966	1,234.2	1,270.0	123.8
1967	1,391.3	1,391.3	131.8
1968	1,377.1	1,322.0	120.2
1969	1,441.5	1,313.2	116.0
1970	1,112.1	956.4	93.2
1971	819.8	675.5	64.5
1972	901.7	720.5	68.2

Continued

Source: U.S., Department of Commerce, Bureau of the Census, *Census of Manufactures* and *Annual Survey of Manufactures* (Washington, D.C.: Government Printing Office, for appropriate years).

**Value of purchasing power of consumer dollar used to convert current dollars to constant dollars is shown in Table A-3.

Table A-8 *(Continued)*

BENCH-MARK DATA FOR TABLE A-7

	Computers and Related Machines (SIC 3573 and 3574)				
	Value Added by Manufacture (in Millions of Dollars)			Man-Hours (in Millions)	
Year	Original	1958 Adjustment (Less 2.0 Percent)	Constant Dollars**	Original	1958 Adjustment (Less 2.0 Percent)
1947	$ 231.2	$226.6	$ 338.8	85.4	83.7
1949	230.4	225.8	316.3	71.9	70.4
1950	260.7	255.5	354.4	73.1	71.6
1951	310.1	303.9	390.5	85.1	83.4
1952	342.4	335.5	422.1	86.6	84.9
1953	384.6	376.9	470.4	91.4	89.5
1954	405.4	397.3	493.4	88.5	86.7
1955	407.7	399.6	498.3	92.5	90.6
1956	591.4	579.6	712.3	108.0	105.8
1957	658.6	645.4	765.4	113.8	111.5
1958	579.1		668.9	98.5	
1959	722.5		827.3	98.4	
1960	822.0		926.4	120.3	
1961	915.6		1,021.8	120.3	
1962	1,025.8		1,132.5	118.1	
1963	1,101.5		1,201.7	119.5	
1964	1,762.6		1,896.6	128.8	
1965	2,146.6		2,271.1	141.6	
1966	2,827.7		2,909.7	169.6	
1967	2,444.6		2,444.6	162.8	
1968	2,800.7		2,688.7	171.4	
1969	3,410.4		3,106.9	199.7	
1970	3,563.8		3,064.9	186.6	
1971	3,021.8		2,490.0	149.7	
1972	3,817.8		3,050.4	163.7	

Continued

Source: U.S., Department of Commerce, Bureau of the Census, *Census of Manufactures* and *Annual Survey of Manufactures (Washington, D.C.: Government Printing Office, for appropriate years).*

**Value of purchasing power of consumer dollar used to convert current dollars to constant dollars is shown in Table A-3.

Table A-8 *(Continued)*
BENCH-MARK DATA FOR TABLE A-7

	Household Laundry Equipment (SIC 3633)		
	Value Added by Manufacture (in Millions of Dollars)		Man-Hours (in Millions)
Year	Original	Constant Dollars**	Original
1947	$161.8	$241.9	48.6
1949	120.0	168.1	29.0
1950	165.8	230.0	37.4
1951	173.3	222.7	36.3
1952	195.0	245.3	38.1
1953	213.4	266.3	42.5
1954	190.7	236.8	32.5
1955	241.6	301.3	36.6
1956	271.2	333.3	38.2
1957	238.3	282.6	31.4
1958	324.5	374.8	35.8
1959	347.4	397.8	37.5
1960	279.1	314.5	32.9
1961	313.8	350.2	30.4
1962	346.4	382.4	30.6
1963	353.9	386.1	28.8
1964	379.7	408.6	32.8
1965	392.9	415.7	32.1
1966	415.8	427.9	36.0
1967	408.4	408.4	33.7
1968	500.9	480.9	36.8
1969	538.1	490.2	36.5
1970	504.3	433.7	33.6
1971	593.6	489.1	33.1
1972	684.7	547.1	36.9

Continued

Source: U.S., Department of Commerce, Bureau of the Census, *Census of Manufactures* and *Annual Survey of Manufactures* (Washington, D.C.: Government Printing Office, for appropriate years).

**Value of purchasing power of consumer dollar used to convert current dollars to constant dollars is shown in Table A-3.

Table A-8 *(Continued)*
BENCH-MARK DATA FOR TABLE A-7

	Motor Vehicles and Parts (SIC 3711, 3712, 3714)						
	Value Added by Manufactures (in Millions of Dollars)				Man-Hours (in Millions)		
Year	Original	1958 Adjustment (Plus 2.7 Percent)	1967 Adjustment (Less 7.4 Percent)	Constant Dollars**	Original	1958 Adjustment (Plus 3.1 Percent)	1967 Adjustment (Less 8.1 Percent)
1947	$ 3,544.9	$3,640.6	$ 3,371.2	$ 5,039.9	$1,102.5	$1,136.7	$1,044.6
1949	4,583.4	4,707.1	4,358.8	6,106.7	1,103.2	1,137.4	1,045.3
1950	5,919.9	6,079.7	5,629.8	7,808.5	1,281.8	1,321.5	1,214.5
1951	5,635.9	5,788.0	5,359.7	6,887.2	1,217.3	1,255.0	1,153.3
1952	5,819.3	5,976.4	5,534.1	6,961.9	1,147.9	1,183.5	1,087.6
1953	6,940.9	7,128.4	6,600.9	8,237.9	1,385.9	1,428.8	1,313.1
1954	5,901.4	6,060.8	5,612.3	6,970.5	1,125.9	1,160.8	1,066.8
1955	9,432.1	9,686.8	8,970.0	11,185.6	1,436.2	1,480.7	1,360.8
1956	7,628.0	7,834.0	7,254.3	8,915.5	1,145.7	1,181.2	1,085.5
1957	8,291.3	8,515.1	7,885.0	9,351.6	1,129.1	1,164.1	1,069.8
1958	6,504.5		6,023.2	6,956.8	858.4		788.9
1959	8,915.4		8,255.7	9,432.8	1,058.0		972.3
1960	9,717.4		8,998.3	10,141.1	1,131.5		1,039.8
1961	8,541.7		7,909.6	8,827.1	945.1		868.5
1962	11,110.9		10,288.7	11,358.7	1,110.2		1,020.3
1963	11,433.9			12,474.4	1,098.1		1,192.6
1964	11,975.3			12,885.4	1,122.9		1,226.4
1965	14,691.3			15,543.4	1,311.0		1,426.8
1966	14,476.6			14,896.4	1,298.0		1,412.7
1967	13,065.6			13,065.6	1,133.4		
1968	16,575.8			15,012.8	1,306.6		
1969	16,923.3			15,417.1	1,287.8		
1970	13,780.3			11,851.1	1,064.2		
1971	19,795.2			16.311.2	1,204.9		
1972	20,927.4			16,721.0	1,282.3		

Source: U.S., Department of Commerce, Bureau of the Census, *Census of Manufactures* and *Annual Survey of Manufactures* (Washington, D.C.: Government Printing Office, for appropriate years).

**Value of purchasing power of consumer dollar used to convert current dollars to constant dollars is shown in Table A-3.

Table A-9
PORTION OF U.S. TOOL AND DIE INDUSTRY IN MICHIGAN
AS INDICATED BY KEY ECONOMIC STATISTICS: 1972, 1967, 1963, AND 1958

Item	1972			1967			1963			1958		
	Michigan	U.S.	Michigan Portion (Percentage) of U.S.	Michigan	U.S.	Michigan Portion (Percentage) of U.S.	Michigan	U.S.	Michigan Portion (Percentage) of U.S.	Michigan	U.S.	Michigan Portion (Percentage) of U.S.
Establishments	1,205	6,616	18.2	1,256	6,615	19.0	1,092	5,896	18.5	1,097	5,745	19.1
All Employees (1,000's)	25.7	98.1	26.2	32.3	113.6	28.4	25.7	90.9	28.3	24.4	83.3	29.3
Total payroll (millions of dollars)	338.5	1,116.3	30.3	334.8	1,032.8	32.4	222.2	682.7	32.5	183.1	536.6	34.1
Production workers: Man-hours (millions)	46.4	173.5	26.7	62.4	214.2	29.1	50.3	169.3	29.7	42.5	139.9	30.4
Production workers: Wages (millions of dollars)	257.7	846.2	30.5	264.6	800.0	33.1	178.9	543.0	32.9	144.0	415.2	34.7
Value added by manufacture (millions of dollars)	535.1	1,799.4	29.7	518.3	1,646.7	31.5	338.6	1,029.3	32.9	260.6	780.1	33.4
Value of shipments (millions of dollars)	734.2	2,426.7	30.3	697.3	2,202.3	31.7	455.7	1,388.8	32.8	354.4	1,060.6	33.4
New capital expenditures (millions of dollars)	22.8	97.0	23.5	32.6	114.7	28.4	15.0	53.3	28.1	8.7	43.2	20.2

Source: U.S., Department of Commerce, Bureau of the Census, *Census of Manufactures: 1972*, MC72(2)–35C, *Census of Manufactures: 1967*, 35C–9; *Census of Manufactures: 1963*, p. 35C–8; *Census of Manufactures: 1958*, p. 35C–8, and *Census of Manufactures: 1958*, p. 35C–7 (Washington, D.C.: Government Printing Office, appropriate years).

APPENDIX B

MAKE-OR-BUY PROGRAMS OF THE BIG FOUR

Highlights of U.S. automotive manufacturers' make-or-buy programs, as they relate to tools and dies, are summarized in this appendix. The complete record of the detailed statements presented at the 1969 U.S. House of Representatives hearings has been published by the committee.

American Motors. American Motors relies 100 percent on outside companies for special tooling and dies used in the production of component parts. The only component tooling fabrication within our corporation is for normal maintenance and repair.
...More than 85 percent of our tooling costs for the 1970 model cars will be from sources within the United States. Our decision to procure the remainder abroad is due to three basic business considerations which influence our purchasing decisions. These considerations are price, delivery, capability, and quality....Frankly, we prefer sources within the United States for very practical reasons. The delivery time is usually shorter and engineering changes can ordinarily be coordinated with greater facility....When foreign sources present us with material cost reduction opportunities, we feel we cannot overlook the advantages....[1]

Chrysler Corporation. Our expenditure for tools and dies varies from one year to the next, however, principally as a result of changing our own in-house capacity....
Of this expenditure [for special tooling], more than 50 percent of the necessary die work, the great majority of our fixtures, and almost all our special machines were provided by outside sources....

Our experience has proved that it is essential for us to maintain in-house die operations sufficient to meet essential tryout and maintenance requirements to satisfy the demand for new-product security, and to insure that our needs will be adequately met.

We are somewhere near 50 percent in-house die activity at present. We believe this level will prove sufficient....

No tools or dies are imported into the United States by Chrysler Corporation from either Europe or Japan....

I am given a certain allocation of funds in a project to bring out a new model, and we, as a practice, estimate the cost of all the dies, both inside and out, so that we can be sure that we are checking the purchasing department, and in turn, on the in-house tools, the same group of people are checking to make sure that we are manufacturing all of them at what is, you might say, a competitive price....If it can be done less expensively, we will put it on the outside.[2]

Ford Motor Company. Only two of our facilities produce any appreciable amount of new tooling. These facilities are at our Canton, Ohio, forge plant, which for years has produced all of its forging dies, and the plants of our metal stamping division....

We have tooling facilities at each of our five stamping plants, as well as in a separate metal stamping division tool and die plant in Dearborn....

The proportion of body die business placed with outside sources for model years 1959–65 has ranged from 59 to 70 percent.

Our tool and die capacity was increased about 10 percent in 1965 when a new stamping plant was phased in at Woodhaven, Michigan....Our overall effective tool and die capacity also has been increased somewhat in recent years through technological improvements....

Subcommittee question: I assume you make a cost analysis on a particular job [tool or die] to determine whether it should be contracted rather than performed in-house?

Reply of company representative: No, our procedure is not to do this on a cost basis. It is based on our in-house capability of making certain of the dies and the capability on the outside of making the dies, meeting our delivery, and otherwise fitting into our total schedule. We do not put the outside shops and our inside shops in direct competition.[3]

General Motors Corporation. General Motors has an obvious interest and obligation to maintain stability in the work schedules of skilled tradesmen on our payrolls. There would be serious labor relations consequences if it were necessary to place large numbers of our employees on layoff in order to assign a specific amount of die-construction work to outside shops. We have assured the unions which represent our employees that it is our intent and desire to use our own skilled trades employees to do the kind of work that they have customarily done in our plants in the past, to the extent that it is practicable and economical to do so. . . .

Thus, the chart shows that slightly more than half of our total requirements for passenger car sheet metal die construction has been performed in-house, while slightly under half of the workload has been given to outside job shops. I would again like to emphasize at this point that we are discussing in connection with this chart die construction for outer and inner panels and other sheet metal parts for General Motors passenger cars. This represents approximately 75 to 80 percent of our average annual die-construction needs for all purposes. . . .

The chart also shows that in the 1967, 1968, and 1969 model years our total die-construction requirements were above the ten-year average. Most of these increased requirements went to the outside job shops, as the GM in-house workloads fluctuated only slightly above the ten-year average. . . .

There are several factors—well-known to all within the industry—which must figure in reaching a sound business judgment on establishing added die-construction capacity. These factors are not unique to General Motors, and I would like to review them.

As noted earlier in connection with our Kalamazoo plant, die-construction capability is an integral part of a stamping facility. This is a general practice in the industry. Beyond this, management must consider a variety of factors in estimating the most efficient capacity level. At one extreme, the point at which the auto manufacturer has no die construction or repair capacity, total potential operating costs are at a maximum. The other extreme, 100-percent capacity internally to produce dies, is equally inefficient. Because of the cyclical nature of tooling requirements, the manufacturer would find himself in the latter instance with unused capacity in all but peak years. Therefore, the optimum level is someplace between these points.

In measuring the optimum point, we consider the fact that we have a continuing need for die repair, die maintenance, [and] die rework for modifications for subsequent model years and engineering changes in current dies and die tryout. We gain efficiency and minimize cost if we have capacity to accommodate these needs.

Also, we must consider timing—whether outside shops would be able to meet tight deadlines or difficult leadtime requirements. As noted in the discussion of Chevrolet–Flint—we must also consider whether outside tooling facilities, adequate to do certain jobs, are available.[4]

NOTES

1. *Hearings* before the Subcommittee on Special Small Business Problems of the Select Committee on Small Business pursuant to House Resolution 66. 91st Cong., 1st Sess. (1969), pp. 169–70.

2. *Ibid.*, pp. 122, 123, and 135–36.

3. *Ibid.*, pp. 109 and 111–12.

4. *Ibid.*, pp. 137, 139–140, and 141.